AIR DISASTERS

AIR DISASTERS
DRAMATIC BLACK BOX
FLIGHT RECORDINGS

MALCOLM MacPHERSON

Collins

Collins
A division of HarperCollins*Publishers*
77–85 Fulham Palace Road,
Hammersmith, London W6 8JB

www.harpercollins.co.uk

First published by Collins 2008
1

British Library Cataloguing in Publication Data
A catalogue record for this book is
available from the British Library

Mixed Sources

Product group from well-managed
forests and other controlled sources
www.fsc.org Cert no. SW-COC-1806
© 1996 Forest Stewardship Council

FSC is a non-profit international organisation established to promote the
responsible management of the world's forests. Products carrying the FSC
label are independently certified to assure consumers that they come
from forests that are managed to meet the social, economic and
ecological needs of present and future generations.

Find out more about HarperCollins and the environment at
www.harpercollins.co.uk/green

ISBN: 978–0–00–728089–6

Set in Sabon by
Rowland Phototypesetting Ltd,
Bury St Edmunds, Suffolk

Printed and bound in Great Britain by
Clays Ltd, St Ives plc

CONTENTS

INTRODUCTION

Before the publication in 1984 of my book *The Black Box* I had called at the offices of the US National Transportation Safety Board (NTSB) in Washington, DC, on the unlikely chance that the Board made available to the public transcripts of commercial air transportation cockpit voice recordings (CVRs), which I knew about from commercial pilot friends who read them out of personal interest and had handed them on to me. My request that day surprised the Public Affairs officer who said that no member of the public had asked to see them before. He wondered out loud what anyone would want with them, as he led me over to thirty or forty transcripts that someone in the office had patiently typed out, copied and stacked up against filing cabinets, looking as if they were ready to be thrown out.

I told him that they fascinated me in a way I could not describe even to myself but their hold on me lasted long after I read them. Maybe they would interest other like-minded people if they were made available in book form. Come to think of it, he told me, he read the transcripts, too, and not for strictly professional reasons. He could not articulate his interest either, but we sat on the office floor and, from memory, he chose twenty of the transcripts that he thought would be of most interest to me. Before long I staggered out of the building under the weight of at least a score of the transcripts. As soon as possible I set about turning the pages into a form that would make sense to readers like me with no specialized knowledge of flying or of the mechanics of commercial aircraft and aviation. *The Black Box*, as a specialized book category, was born.

Now, nearly thirty years on, the aviation industry has changed remarkably, as anyone who flies knows. I must say that most of these changes are for the worst from the passengers' perspective. Today, already cramped spaces in aeroplanes are smaller, meals are worse or in some cases nonexistent, delays are longer, cancelled flights are stacked up higher, rude and indifferent behaviour by officials seems almost routine, and more people like you and me want only a quick and merciful end to the experience of flying between home and destination that does not involve a crash. In 2007 passengers' complaints to the US Department of Transportation increased by 60 per cent, which means that, for instance, a typical passenger from Washington Dulles Airport was angry enough with the service to complain once every day and a half. In an editorial in the spring of 2008 the *Washington Post* had this to say: 'Air travel has gotten so bad these days that going to the airport requires either an exercise in sadomasochism or an abiding faith that everything will be okay. That faith seems to be shattered daily.' Beneath the surface passengers are seething; to tell the truth, flying is no longer the convenient way to travel, but in America it is the *only* way, at least with gasoline prices inching up, train services starved of federal funds and bus travel nearly nonexistent. Bad as it is, we have no other choice but to fly.

That said, in one very important – the *most* important – respect, changes over the years in commercial air travel have been excellent for passengers: whenever we set out to reach a destination by air, we arrive. In other words, though flying might be miserable it is safe, pure and simple, and, remarkably, it is getting more so all the time. As if air safety were a warp zone of science fiction, safety managers for airlines and the government are today reaching beyond known weak links into the realm of 'what might conceivably happen' and are making changes for safety before trouble happens, so that passengers never have to experience incidents that did not have a chance to develop.

Indeed, there has been no major aircraft crash in the last seven years in the USA. That's safety to bank on. Indeed, 2007 was a typically excellent year. With 4.65 *billion* air passengers travelling worldwide (769 million on aeroplanes based in the US), 965 people died of all causes in 136 incidents, which were 28 fewer than in 2006, with a 25 per cent decrease in fatalities (and none of these, in fact, included commercial aviation). In the United States 95 per cent of transportation fatalities occur on roads and highways. Waterways and railroad lines account for more fatalities than aviation, which weighs in at a mere 1 per cent of the total for all transportation – or roughly eighty times safer than travel on roads. The time can almost be foreseen when compilations similar to the one you are now reading won't exist for lack of enough CVR transcripts to put between covers.

Not surprisingly, the safety trends in the United Kingdom, according to the statistics offered by the Civil Aviation Authority (UK), have run closely in parallel with those in the US. While the number of passengers flying to, from or between UK airports more than quadrupled between 1980 and 2006, from 50 million to 210 million, safety has steadily improved to the point of near statistical perfection. In 2006, 185 million passengers arrived or departed the UK on international flights, while 25 million passengers travelled on domestic flights. And yet fatality rates were less than 1 billion passenger kilometres in all years since 1981 and less than 0.1 per billion since 1990. Within the UK, there was one fatality among airline (and air taxi) passengers reported since 2001 and five among airline crews; UK airlines flying *outside* the British Isles reported no passenger or crew fatalities between 2000 and 2006. (The reason you will find no UK CVR transcripts in this collection is because there have recently been no fatal accidents to record and transcribe, even if the Air Accidents Investigation Branch, part of the Department of Transport, published CVR transcripts or otherwise made them available to the general public.)

The airline industry and government bureaucracies continue to work hard to keep flying safe, aware that fallible human beings will always be involved to create risks. Starting on this new book, I was curious to find out why airline fatalities started to decline, almost precipitously, around 1963. I found that by then the aviation industry had switched over from internal combustion engines to much simpler, and thus safer, jet engines. It was also then that the industry started to build redundancies into aeroplanes beyond just multiple engines. Redundancy, at its simplest level, catches failures. On a two-engine airliner, for instance, if one engine fails on takeoff, the aircraft design

4

calls for the remaining engine to propel the entire weight of the aircraft by itself. Duplication was installed wherever possible.

To this was soon added another safety innovation – the nearly obsessive maintenance of records with which to cross-reference minute and large mechanical and other aircraft failures to forestall other identical (or similar) failures. For instance, when United Airlines DC-10 Flight 232 crashed while attempting to land in Sioux City, Iowa, on 19 July 1989, the NTSB determined post haste that engine failure, and specifically the failure of a single disc in one of the aircraft's three fanjet engines, had caused the crash. Discs in six engines had been forged from the same alloy ingot, and in a matter of days the other five aircraft with engine discs made from that ingot were identified and grounded, thus eliminating the possibility of a second or third engine failure. That kind of thoroughness is unparalleled.

Over the years starting around 1963, pilot training improved as well. All cockpit crews today train in sophisticated, aircraft-specific simulators. Before simulator use, pilots necessarily acquired experience in the aircraft in real time, which clearly prevented exploration of the aircraft at the edge of its performance envelopes and beyond to see what would happen. Now, with simulators, pilots can crash aeroplanes over and over again while sitting inside a building. Not long ago, I flew 'aboard' a simulator of a Chinook military helicopter (MH-47) through an emergency crash/dive that had nearly killed everyone on board in Afghanistan a couple of years earlier. The helicopter pilot in Afghanistan was my pilot in the simulator at a US

Army base, and such was the reality of the experience that even the simulator's technical minders paled; with lights and horns blaring and the cockpit shaking and controls trembling, and, with mountainous Afghan terrain out of the front window, I did not feel that anything was being simulated. I walked away with the reassurance that cockpit crews are the best prepared and trained for any emergency ever.

For decades cockpits were considered authoritarian domains. Like that of a captain on board a ship, the word of the captain in the cockpit was law. Crashes and other incidents occurred when the captain failed to listen to a warning from the co-pilot or first officer. Sometimes, first officers did not even bother to speak up when they saw trouble ahead. Slowly, from around the mid-1980s, cockpit crews began to move away from stratification and develop a new structure in which the cockpit 'team' became the ruling entity. First officers were encouraged to speak up and captains were instructed to pay attention to them, for the sake of everyone's safety. Older pilots who felt uncomfortable in this more democratic setting were asked to resign. The atmosphere improved immeasurably and in terms of safety everyone benefited, including the captain.

At the same time, a vast new array of technology emerged to assist cockpit crews. Technologies that monitor systems, navigate and effectively fly the aeroplane became commonplace in cockpits. And while these tools helped cockpit crews, the individual in the cockpit did not succumb to a natural inclination to rely on technology. Crews today are trained never to let the aeroplane be flown by computers against the pilot's common sense and experience.

The weak link in safety is man, but that link can also be a last hope.

Taking safety an extra step into a realm where any unknown threat can be isolated and examined before an accident occurs, airlines today are making full use of a technology called Flight Operations Quality Assurance (FOQA), first developed in Europe (and known there as FDM or flight data management). FOQA uses flight data stored in quick-access recorders; the data are analyzed every dozen or so flights. Essentially, FOQA enables airlines to search the data for any events or trends that might signal a conflict with normal, or standard, operating procedures. It affords airlines a real-time audit of what's going on. This can lead to changes that make flying safer. For example, United Airlines routinely examined FOQA data on its aircraft flying to Mexico City, where it noticed that oddly fast approaches to landing had the potential, at least, of causing runway overrun accidents. United wanted to know what was causing the faster approaches, and FOQA told them. Their aircraft were typically being told to turn early, before a designated intersection. That early turn put the aircraft higher – and thus faster – on the approach. Corrections were made and a potential problem was corrected before it actually arose.

FOQA has also helped to eliminate costly maintenance problems. In one recent example pilots were reporting that they did not know why they were exceeding maximum speeds for deploying wing flaps. The airline analyzed the FOQA data on their aircraft. They reported to these pilots that their speeds were on average only 1 knot or less over maximum. If the airline had gone on the word of the pilots it would have had no other choice but to take the aircraft out of service and disassemble the wing. With FOQA, no such response was necessary and costs were avoided and money saved. Other savings were found through FOQA in gasoline conservation. Conserving only one gallon of AVGAS over a flight by reaching higher altitudes sooner can save an airline tens of millions of dollars a year.

With the recent rapid growth of demand for airline services, new safety concerns will always occur in spite of efforts to anticipate them. One such current concern is called 'runway incursion'.

While concentrating on making air travel safer in the air than on the ground, airports today are experiencing serious congestion problems on runways and taxiways – specifically, incursions. 'There are more planes but not more cement', one person associated with analyzing the issue told me. Congestion leads to incursions, which can lead to crashes and fatalities on the ground. The rate of new incursions has alarmed the US Federal Aviation Administration (FAA) and other nations' oversight groups. In response, considerable effort is being made to warn cockpit crews directly of impending incursions. At the present time, warnings routinely pass through ground controllers who analyze the specific warnings before alerting the cockpit crews. Those precious lost minutes can make the difference between life and death. The FAA is also studying the benefits of obvious warning lights on runways and taxiways, similar to the lights at road intersections.

Commuter and regional airlines have come under the FAA's microscope with increased demand for short-hop flights, which feed into major airline hubs. These flights are crashing at rates that worry air safety managers, who have focused their attention on operational crew training of regional airlines. And as more crews receive increased simulator time and intensified qualifying tests and checks, the safety record is bound to improve.

And, finally, to put your minds at rest, there are these astonishing statistics:

Odds of being on an airline flight which results in at least one fatality	Odds of being killed on a single airline flight
Top 25 airlines with the best records 1 in 6.06 million	Top 25 airlines with the best records 1 in 10.46 million
Bottom 25 with the worst records 1 in 546,011	Bottom 25 with the worst records 1 in 723,819

But enough about air safety. No reader of this book will have got this far wanting to know more about flight safety. Let's admit the obvious. This book is unabashedly about air *un*safety. The mention of air safety in an *un*safety book is a fig leaf for the real reason for publishing CVR transcripts. Over the years I have been editing books such as this some readers have told me that the CVR transcripts actually help calm their nerves, though I do not understand how, since most of the incidents recorded here end with dead bodies and charred aircraft. Truly, these transcripts should give readers the heebie-jeebies. Maybe the claims that nerves are calmed stem from rehearsing a disaster at a distance, imagining what we might do or not do in these same dire straits; the CVR transcripts give some readers the illusion of being in control, when, as passengers, we have absolutely no control over whether we live or die;

9

and we know it. This might seem obsessive, but the thinking must go that if a reader follows these disasters often enough by rereading the CVR transcripts, when (and if) the time comes to experience one such incident for real he or she will be ready. Maybe that is true for some people. They have already been there, so to speak. But I also suspect that more readers follow the transcripts for their drama, as I did at the start. It is undeniable that they make riveting reading, because they document real life-and-death events as they unfold minute by minute from a spectacular angle. We can follow the activities, emotions and voices of cockpit crews from the instant something goes awry to a final outcome. And what can be more dramatic than for-real death or salvation? All drama, whether portrayed as fiction or fact, is necessarily voyeuristic. And what can be more intrusive than peeking from behind the curtain at the last frenzied, intimate moments in another human being's life?

*Un*safety will be with us in the air for a long time to come. Flying in North America and Europe may have reached a point of statistical perfection but we will still have the Third World, which is where airlines are crashing today.

The imbalance in safety between different parts of the world is stark.

In March 2007, Russian Airlines UTair Flight 471, a Tupolev Tu-134, which crashed while attempting to land at Samara's Kurumoch Airport, in Russia, killing six of the fifty-seven passengers on board, was *only one of two fatal commercial passenger aircraft accidents that did not occur in the Third World or involve an aircraft registered there*. (The worst aviation disaster of 2007 was the crash of the

Brazilian TAM Linhas Aéreas Flight 3054, an Airbus A320 that overran the runway at Congonhas-São Paulo International Airport in Brazil, killing 187 on board and 12 on the ground.) In the Samara incident the aircraft was a Tupolev; in case you did not already know, boarding any Tupolev anywhere, flown by any airline, whatever its destination, is guaranteed to be the thrill of a lifetime. My wife and I flew in one a few years ago from the Bahamas to Havana, Cuba. The subsequent vacation, the cigars, music, food and sun and rum were just a pleasant afterthought to the joy of having landed alive.

As in so many other aspects, Africa, in terms of air safety, has become a special case, with the European Union banning most non-national African airlines from landing at EU members' airports. Last year the Congo saw more fatal commercial air crashes than any other country. Four cargo aircraft and two Let 41 passenger flights suffered fatal crashes. One of these accidents involved an Africa One plane which came down in Kinshasa, killing so many people on the ground that no precise number of casualties was ever given. The downward trend in the safety of African airlines is long and continuous. Even as far back as the early 1970s, flying in Africa required a cavalier attitude. I remember when, based in Kenya with *Newsweek* magazine in 1973, I was aboard an Air Zaire flight from Kinshasa to Nairobi, and, to my surprise, a beautiful young woman who had boarded the Boeing 707 aircraft in Kinshasa simply vanished soon after takeoff. I know because I looked for her. A week later I ran into her at a Nairobi dinner party. She laughed as she recalled the thrill of the flight, telling me that, although unqualified, she had piloted long stretches between Bujumbura and Entebbe while sitting in the captain's lap swilling goblets of champagne.

To change this direction in African air safety, the American NTSB and others are working 'aggressively' to help African nations. At first glance the continent would appear to extend beyond the NTSB's mandate, and indeed it does . . . and yet doesn't. African governments and private airlines based in Africa buy and fly American-made aeroplanes and helicopters, which gives the NTSB an inherent interest. 'For commercial purposes, we don't want them crashing Boeings', a member of the NTSB told me. 'We want them *buying* Boeings. If there are crashes, the African governments or airlines might say that the plane was no good. "Next time we'll buy Airbus".' This was the line the Egyptian government took when one of EgyptAir's pilots committed suicide in October 1999 by crashing a Boeing 767 into the Atlantic Ocean. The Egyptian government alleged that the 767's flap system was to blame. It wasn't. 'That was not good for business,' the same NTSB representative told me.

Air safety in Africa can also have serious political ramifications. In August 2005, John Garang, the newly sworn in vice president of the Sudan, was killed when his helicopter crashed during an official trip to Uganda. Soon after, the BBC reported 'large-scale' rioting in the Sudanese capital Khartoum, with supporters of Mr Garang battling armed police. They inferred from the news that Sudanese enemies in positions of authority had ordered his killing. This suspicion was perhaps inspired by the shooting down of a plane in 1994 that was carrying Rwanda's President Habyarimana, an incident that served as a flashpoint for the subsequent genocide. Immediately after the 2005 crash in Uganda, the US Department of State dispatched one of the NTSB's seasoned investigators, Dennis Jones, to the crash scene. His conclusion that foul play was not involved may have prevented a bloody civil war.

For readers unfamiliar with CVR transcripts, an explanation is in order. The transcripts are taken from recordings of sounds of interest to investigators after crashes. These are inter-cockpit voices, engine noises, stall warnings, landing gear extension and retraction, and all sorts of other clicks and pops. Investigators can often determine from these noises parameters such as engine rpm, systems failures, speed, and the time at which certain events occur. The CVR tapes also record communications with air traffic control, automated radio weather briefings and conversation between the pilots and ground or cabin crew.

All over the world these recordings are contained in boxes carried in the parts of commercial aircraft most likely to survive a crash, such as the tails. These boxes are known colloquially as black boxes; there is one for cockpit voice recordings (CVR) and another for flight data recordings (FDR). In the cockpit, the crews' voices and other sounds are detected by 'cockpit area microphones', or CAM, usually located on the overhead instrument panel between the two pilots. The older analogue CVR units use a quarter-inch magnetic tape as a storage medium on a thirty-minute self-erasing loop. Newer models use digital technology and memory chips for up to two hours of self-erasing recordings. The boxes contain an underwater locator beacon (ULB), which activates a 'pinger' when the recorder is submerged in water and transmits an acoustical signal on 37.5 KHz that a special receiver can detect at depths of 14,000 feet. The boxes can sustain a crushing impact of 3400 gravities of force. One even sustained 9000 gravities after the crash in 1987 of a hijacked Pacific Southwest Airlines flight in California.

The cockpit voice recorder (CVR)

After an accident occurs and the black boxes are located at the accident site they are pulled from the wreckage and quickly delivered to a laboratory where they are opened and examined. In America that examining agency, the NTSB, is located at L'Enfant Place, Washington, DC. To listen to the tapes a CVR committee is formed from the representatives of the airlines, the manufacturers of the aircraft and of its engines, pilots' unions and the NTSB. This committee compiles a written transcript of the CVR to be used during the investigation, and examples of such transcripts, edited by the NTSB investigators, are what you are reading, for the most part, in this book. FAA's air traffic control tapes, with their associated time codes, are used to help determine the local standard time of one or more events during the accident sequence. The transcripts, containing all pertinent parts of the recording, are edited. Anything other than factual information is removed in the knowledge that the transcripts very often contain highly personal and sensitive verbal communications inside the cockpit.

Cockpit voice recorder	
Time recorded	30 minutes continuous, 2 hours for solid-state digital units
Number of channels	4
Impact tolerance	3400Gs/6.5 ms
Fire resistance	1100 deg C/30 min
Water-pressure resistance	submerged 20,000 feet
Underwater locator beacon (ULB)	37.5 KHz; battery has shelf life of six years or more, with thirty-day operation capability upon activation

I have chosen the following twenty-one transcripts on the basis of their variety and drama. I make no apologies for what to some might seem ghoulish. In editing these transcripts for publication I have not tried to 'characterize' the crew members whose voices are taken directly off the CVR tapes. I do not want these transcripts to read like an airport novel. Whether the captain of the downed aircraft was kind to animals, was married with children, etc – none of this seems to me to be relevant in an accident; the same goes for the passengers whose lives are equally unknown to me. I have tried to give readers a context – of weather, time, numbers of passengers, sights and sound. I have edited some of the crews' dialogue for clarity and I have qualified some of the pilots' jargon with bracketed definitions that laymen better understand. I want readers to know that I am not a pilot. I have never been a pilot. I have not edited this book for pilots or for other aviation experts who will almost certainly be better served reading the original versions of these transcripts.

Finally, readers might be advised to imagine themselves, rather than sitting aboard the aeroplanes mentioned in the following pages, tuned to a radio and overhearing the sounds as they happened, and events as they unfolded. Even if you are not able to visualize everything, I know you will agree with me that these transcripts are as dramatic reading as you are likely to find, because they are minute-by-minute, unvarnished accounts of what actually occurred.

Malcolm MacPherson

COLORADO SPRINGS, Colorado, USA

3 March 1991

On 3 March 1991 a United Airlines (UAL) Boeing 737, registration number N999UA, operating as Flight 585, was on a scheduled passenger flight from Denver to Colorado Springs, Colorado. The weather in Colorado Springs was clear, with visibility 100 miles, temperature 49 degrees Fahrenheit, dew point 9 degrees Fahrenheit, winds 330 degrees at 23 knots, gusts to 33 knots, and with cumulus clouds over the mountains to the north-west. The captain was flying the aeroplane and the first officer was working the radio transmissions. The plane was scheduled to arrive in Colorado Springs at 9.46 a.m.

09:41:20 CAPTAIN: Twenty-five flaps.
09:41:23 TOWER: United 585 after landing hold short of runway 30 for departing traffic on runway . . . 30.
09:41:25 [Sound similar to that of an engine power increase]
09:41:30 CAPTAIN: Starting on down.
09:41:31 FIRST OFFICER: We'll hold short of [runway 30], United 585. That's all the way to the end of our runway not . . . doesn't mean a thing.
09:41:39 CAPTAIN: No problem.

09:41:51 [Sound similar to that of stabilizer trim actuation]

09:42:08 FIRST OFFICER: The marker's identified. Now it's really weak.

09:42:11 CAPTAIN: No problem.

09:42:29 FIRST OFFICER: [We had a] ten knot change here.

09:42:31 CAPTAIN: Yeah, I know . . . awful lot of power to hold that . . . airspeed.

09:43:01 FIRST OFFICER: Another ten knot gain.

09:43:03 CAPTAIN: Thirty flaps.

09:43:08 FIRST OFFICER: Wow.

09:43:09 [Sound similar to that of an engine power reduction]

09:43:28.2 FIRST OFFICER: We're at a thousand feet. Oh, God [the aircraft flips over] –

09:43:33.5 CAPTAIN: Fifteen flaps.

09:43:34 FIRST OFFICER: Fifteen. Oh.

09:43:34.7 CAPTAIN: Oh! [Loud exclamation]

09:43:35.5 [Click sound similar to that of a flap lever actuation]

09:43:35.7 CAPTAIN: Fuck.

09:43:36.1 [Click sound similar to that of a flap lever actuation]

09:43:36.5 CAPTAIN: No! [Very loud]

09:43:37:4 [Click sound similar to that of a flap lever actuation]

09:43:38.4 FIRST OFFICER: Oh, my God . . . Oh, my God! [A scream]

09:43:40.5 CAPTAIN: Oh, no. [Exclaimed loudly]

09:43:41.5 [Sound of impact]

Numerous witnesses reported that shortly after completing its turn onto the final approach to runway 30 at Colorado Springs Airport at about 9.44 a.m., the aeroplane rolled steadily to the right and pitched nose-down until it reached a nearly vertical attitude before hitting the ground in an area known as Widefield Park. The aeroplane impacted relatively flat terrain 3.47 nautical miles south of the south end of runway 30 and .17 nautical miles to the east of the extended centreline of runway 30 at the airport. Everyone on board the flight (the two flight crew members, three flight attendants and twenty passengers) received fatal injuries. The plane was destroyed by impact forces and post-crash fire.

More than sixty witnesses were interviewed during the initial field phase of the NTSB's investigation and more than a hundred other witnesses came forward during a follow-up visit to the accident site about a year later. The majority of the witnesses indicated that, although the aeroplane was flying at an altitude that was lower than they were accustomed to seeing, it appeared to be operating normally until it suddenly rolled to the right and descended into the ground. Many witnesses reported that the aeroplane rolled wings level momentarily (as it lined up with the runway) and that it rolled to the right until it was inverted with the nose nearly straight down.

Some of the witnesses saw the nose rise during the initiation of the right roll. One elderly couple, reportedly walking through Widefield Park at the time of the accident, stated to another witness that a liquid substance from the aeroplane fell onto their clothing which 'smelled very bad'. Repeated efforts to find and interview this couple have been unsuccessful. One witness, who was about six miles west of the accident site, reported seeing several rotor clouds (the rotor cloud is a form of lee eddy, often associated with extreme turbulence) in the area of the accident, ten to fifteen minutes before the crash. That witness said that the rotor clouds were accompanied by thin, wispy condensation. Another person, who passed west of the accident site between 8.30 and 9.00, reported seeing 'torn wispy clouds' in the area of the accident.

On 8 December 1992 the NTSB issued a final report on the accident. The Safety Board concluded that it 'could not identify conclusive evidence to explain the loss of United Airlines flight 585'. In its statement as to the probable cause of the accident, the Board indicated that it considered the two most likely explanations for the sudden uncontrollable upset to be a malfunction of the aeroplane's directional control system or an encounter with an unusually severe atmospheric disturbance.

MEMPHIS, Tennessee, USA
7 April 1994

FedEx Flight 705 was scheduled to depart the company's home base of Memphis, Tennessee, for a routine flight to San Jose, California, at a little after 3.00 in the afternoon. The captain, first officer and flight engineer boarded the aircraft to find that another FedEx employee, who was not identified as someone who would fly with them that day, had already come on board. He was sitting at the flight engineer's station initiating pre-flight procedures. A 'jump-seat' employee may not interfere with flight operations, according to the rules and regulations of the company, but the boarding flight crew members said nothing and the stranger gave up his seat to the flight engineer before strapping himself into a jump seat in the cockpit. He carried with him a guitar case, which contained two claw hammers, two sledgehammers, a knife and a scuba diver's speargun.

The jump-seat employee, later identified as Auburn Calloway, came aboard Flight 705 that afternoon in a state of desperation. His career with FedEx, and, indeed, with any other freight company or airline, was in jeopardy. Federal Express had uncovered a series of irregularities and outright falsifications in both his original employment application and in hundreds of hours of flight records. The company had ordered him to appear at a disciplinary hearing in the upcoming week. The likely outcome of such a hearing was the termination of his employment and the loss of his FAA flight certification. He decided to kill himself and to make the suicide look like an accident, thus providing insurance money for his family. Moreover, by crashing one of their aeroplanes, he would punish FedEx for singling him out. Calloway's plan was carefully devised. He did not want to use guns to disable the cockpit crew, knowing that any subsequent investigation would uncover bullet wounds. He wanted to make the deaths of the crew look as if they had been as the result of an air crash. He bought death and dismemberment insurance and planned to bludgeon to death the crew of Flight 705, then crash the DC-10 into the terminal of the Memphis base.

After takeoff the captain heard a struggle. He turned to see both his crew mates slumped in their seats, injured terribly, and a blood-soaked Calloway moving towards him. Calloway swung wildly at the captain and, although his blows landed, some were deflected. The plane lurched as the captain defended himself desperately. Then the first officer and flight engineer recovered sufficiently to fight back. Calloway swung a hammer, inflicting further injuries, but the cockpit crew did not give up and Calloway retreated to the rear of the cockpit. The crew did not have time to radio Memphis before Calloway brought out the speargun. He told the crew, 'Sit down! Sit down! This is a real gun, and I'll kill you.'

The flight engineer was bleeding from wounds to his face and head. He could not see Calloway clearly, but he could see the speargun, its barbed steel shaft inches from his face. He grabbed the weapon and threw himself at his assailant. The captain helped him subdue Calloway while the first officer struggled to regain control of the aircraft. The first officer could not use his right arm and the blows to his head had caused near paralysis to parts of his body. He pulled the control yoke back against his chest, and rolled the DC-10 to the left into a barrel roll at nearly 400 miles per hour. The captain and flight engineer shouted, 'Get him! Get him', as the three struggled with Calloway. They fell into the galley area with the movement of the aeroplane, at one moment weightless, at the next being pressed down upon by three times their weight in gravitational forces. The aircraft turned upside down at 19,700 feet.

The first officer threw the aircraft into a series of violent manoeuvres to keep Calloway off balance. He threw the yoke forward and sent the plane into a vertical dive. The throttle controls, located to his right, were pressed forward to their stops, and he could not reach them with his injured right hand. The DC-10 accelerated past 500 miles per hour, then past the instruments' capacity to register the speeds. Flight 705 was experiencing velocity stresses that the airframe was not designed to sustain. The first officer pulled the aircraft out of the dive, and then reached across the yoke with his left hand to cut the speed. Then he called Memphis.

Flight 705 turned back for Memphis and was cleared for any runway. No one on the ground knew what had happened except for an 'attack' reported by the first officer when he had declared an emergency.

Once Flight 705 landed and came to a stop, a paramedic boarded the aeroplane. The captain and flight engineer were still struggling with Calloway on the floor, while the first officer sat trembling in the co-pilot's seat. Calloway was handcuffed and hauled away.

CAPTAIN: I can't believe it, what a goatrope [mess]. What aeroplane number is this?

FIRST OFFICER: It's, uh, 306.

CAPTAIN: Okay.

FIRST OFFICER: We can use auto throttles. [Laughter]

FIRST OFFICER: Express 705 cleared for takeoff. Lights if you want 'em, I mean clocks if you want 'em, lights are coming on, we'll get the vertical speed wheel here in a minute.

CAPTAIN: How's the checklist look?

FLIGHT ENGINEER: Once the flight guidance has been set, we'll be complete.

FIRST OFFICER: All right, er, it's set.

FLIGHT ENGINEER: All right, before takeoff is complete.

FIRST OFFICER: Okay.

CAPTAIN: Your aeroplane.

FIRST OFFICER: I have the aeroplane, set standard power, please, before they change their mind.

CAPTAIN: Power is set.

FIRST OFFICER: Okay.

CAPTAIN: Eighty knots.

FIRST OFFICER: That checks.

CAPTAIN: V one, rotate, positive rate.

FIRST OFFICER: Gear up, please, IAS hold if you can.

CAPTAIN: Right 280, 275 radial outbound, Express 705.

FIRST OFFICER: Check. I don't think you got out.

CAPTAIN: 275 radial outbound, Express 705.

FIRST OFFICER: Okay, climb, well, almost there. Climb power.

CAPTAIN: Express 705, two thousand five for six thousand, Express 705.

FIRST OFFICER: Want CMS?

CAPTAIN: Well, we appear to be safely airborne.

FIRST OFFICER: I was starting to wonder about it.

CAPTAIN: Vertical speed to 1000 feet per minute, please, or thereabouts.

FIRST OFFICER: What a goatrope, what a goatrope!

CAPTAIN: Pardon me?

FIRST OFFICER: What a goatrope back there, jeez!
[Laughter]

FIRST OFFICER: Slats retract. After takeoff checklist [complete]. Let's get out of here.

FLIGHT ENGINEER: Down to the line on the, after takeoff.

CAPTAIN: All right, okay.

FIRST OFFICER: I'll come over here to get that radial.

CAPTAIN: 127.22 for Express 705. Express 705 . . . nine nine. Express 705 is at ten three for 16,000, one six thousand. Express 705 on 27.4 They're out to lunch. Excuse me, but have you worked at our ramp control?
[Laughter]

CAPTAIN: Victor 1 Victor.

FIRST OFFICER: Whew, I got a, oh boy, stay with us.

CAPTAIN: Taxi, Victor 1 Victor . . .

FIRST OFFICER: That's Victor to one Victor, Victor one to Victor. Gee whiz. Well, that's 23.0.

CAPTAIN: Flight level 230 for Express 705. 27.4 Express 704 on 33.0 Leaving one six direct Razorback Express 705.

FIRST OFFICER: Zero one, okay. Here we go.

CAPTAIN: That's Crowley's Ridge, you know about Crowley's Ridge?

FIRST OFFICER: Naw, naw.

CAPTAIN: That's it right there.

FIRST OFFICER: All, all this area right here?

CAPTAIN: See these trees?

FIRST OFFICER: Yeah.

CAPTAIN: That's a natural fault line.

FIRST OFFICER: Oh, this is the New Madrid, uh . . .

CAPTAIN: Well, it's part of it, yeah, but it's much higher in elevation and the . . . climate is different. You drive in Arkansas, you drive right over it.

FIRST OFFICER: Well, I . . .

CAPTAIN: You see all those trees there, that's it.

FIRST OFFICER: I know it, but I wonder about that. You go, Wynne, and all the, you know, stuff over here, you know, where it's flat and you cross over that and I wondered about that. That's not part of the no vaculight uplift and all that, that's where? That's further west, isn't it?

CAPTAIN: Yeah.

FLIGHT ENGINEER: Altimeters.

FIRST OFFICER: Nines and twos here.

FLIGHT ENGINEER: After takeoff is complete.

FIRST OFFICER: Do you, uh, live over in Arkansas, Dave, or . . . ?

CAPTAIN: Naw, I live in Fisherville.

FIRST OFFICER: Aw, Fisherville, great spot.

[Sound of hammer blows striking first officer and flight engineer]

FLIGHT ENGINEER: Ow!

FIRST OFFICER: God!

FIRST OFFICER: Oh, ah, shit.

CAPTAIN: God almighty!

FLIGHT ENGINEER: Ow!

FIRST OFFICER: What the fuck are you doing?

CAPTAIN: God, [groan] God almighty! God, God, God . . .

FIRST OFFICER: Get him, get him, get him.

CAPTAIN: He's going to kill us.

FIRST OFFICER: Get him!

CAPTAIN: Get up, get him!

FLIGHT ENGINEER: I can't, God!

UNIDENTIFIED: Stop! Hold his goddamn . . .

JUMP-SEAT PASSENGER: Sit down, sit down, get back in your seat, this is a real gun, I'll kill ya.

FIRST OFFICER: Get him, get him, get him, get him, get him, get him!

AUTO WARNING HORN: *Bank angle, bank angle.*

FIRST OFFICER: Get him, get him, get him!

JUMP-SEAT PASSENGER: I'm gonna kill you! Hey, hey! I'll kill ya!

AUTO WARNING HORN: *Bank angle, bank angle.*

CAPTAIN: Get him, get him, get him!

AUTO WARNING HORN: *Bank angle, bank angle.*

CAPTAIN: Yeah, get him!

AUTO WARNING HORN: *Bank angle, bank angle.*

FIRST OFFICER: Get him, get him, get him, Andy, I got the aeroplane!

AUTO WARNING HORN: *Bank angle, bank angle.*
FIRST OFFICER: Get him, Andy, get him!
AUTO WARNING HORN: *Bank angle, bank angle.*
[Sound of struggling in background] [Overspeed warning] [Series of clicks]
FIRST OFFICER TO CENTRE: Centre, centre, emergency! Centre, emergency!
AUTO WARNING HORN: *Bank angle, bank angle.*
FIRST OFFICER TO CENTRE: Centre, listen to me! Express 705, I've been wounded. We've had an attempted takeover on board the aeroplane. Give me a vector [direction] please, back to Memphis at this time. Hurry!
AUTO WARNING HORN: *Bank angle, bank angle.*
FIRST OFFICER TO CENTRE: Zero nine five, zero nine five, direct Memphis, get an ambulance and uh, alert the, uh, airport facility!
AUTO WARNING HORN: *Bank angle, bank angle.*
FIRST OFFICER TO CENTRE: Hey, Memphis, you still with me? [Struggling in background]
FIRST OFFICER TO CENTRE: Listen, hey, centre! [Struggling in background]
FIRST OFFICER TO CENTRE: Centre, give me a heading to Memphis!
FIRST OFFICER TO CENTRE: Zero nine zero, roger, this is an emergency . . . ! [Overspeed warning – series of clicks in background] [Struggling in background]
UNIDENTIFIED FROM REAR OF COCKPIT: Let go of it! [Sounds of struggle] Let go of the spear!
FIRST OFFICER TO CENTRE: One zero, thousand, okay, keep me advised, where is Memphis? [Struggling in background]

FIRST OFFICER: Okay, say my direction to Memphis.

FIRST OFFICER: Look, just keep talking to me, okay?

CAPTAIN: Jim!

FIRST OFFICER TO CENTRE: Yeah, we need an ambulance and, uh, we need, uh, armed intervention as well.

FIRST OFFICER TO CENTRE: Down to 5000 feet.

FLIGHT ENGINEER: Put it on autopilot!

FIRST OFFICER TO REAR OF COCKPIT: I've got it!

FLIGHT ENGINEER: Help, the son of a bitch is biting me! [Sounds of struggle]

FIRST OFFICER TO REAR: Andy!

CAPTAIN FROM REAR OF COCKPIT: Put it on autopilot and come back here!

FIRST OFFICER TO REAR OF COCKPIT: Keep him back there guys, I'm flying!

CAPTAIN FROM REAR OF COCKPIT: Put it . . .

FLIGHT ENGINEER FROM REAR OF COCKPIT: Hurry up, Jim . . .

FIRST OFFICER TO CENTRE: Request a single frequency.

FLIGHT ENGINEER FROM REAR: Jim!

FIRST OFFICER TO CENTRE: 19.1.

CAPTAIN FROM REAR: Jim, is it on autopilot?

FIRST OFFICER TO REAR OF COCKPIT: No, I got it.

CAPTAIN FROM REAR: Put it on autopilot and come back here!

FLIGHT ENGINEER FROM REAR: Quick, Jim!

FIRST OFFICER TO REAR: Okay.

CAPTAIN FROM REAR: Hurry, Jim! Come back here now!

FIRST OFFICER TO REAR: Okay . . . wait a minute, I'm coming.

CAPTAIN FROM REAR: Jim, do it now!

FLIGHT ENGINEER FROM REAR: Hurry, hurry!

[First officer exits co-pilot seat and goes to rear of cockpit]

FIRST OFFICER: You move, I'll kill ya!

FIRST OFFICER TO CAPTAIN: Go up and get the aeroplane.

CAPTAIN: I'm going.

FIRST OFFICER: If you can.

CAPTAIN: Are you in control?

FIRST OFFICER: Yeah, can you take this?

[Captain returns to his seat, sounds of him buckling in]

CAPTAIN TO REAR: Jim, are you in control? Jim, are you in control?

FIRST OFFICER FROM REAR: Yes, I'm in control.

CAPTAIN: Memphis, can you hear me?

TOWER: Uh, is this Express 705 heavy?

CAPTAIN: 705, yes.

TOWER: 705 heavy, Memphis, roger, I do hear you. You can proceed direct to Memphis if able. Expect runway niner. The altimeter is 30.29er.

CAPTAIN TO TOWER: You understand we're declaring an emergency. We need security to meet the aeroplane. We'll stop it on the runway if we can.

TOWER: Express 705 heavy, affirmative, all that's been taken care of, that security will be available for, as well as medical assistance.

FIRST OFFICER FROM REAR: Dave!

CAPTAIN TO REAR: Yes!

FIRST OFFICER FROM REAR: Are you okay?

CAPTAIN TO REAR: I'm okay. Are you? Do you have him under control? Talk to me, Jim.

FIRST OFFICER FROM REAR: Huh?

CAPTAIN TO REAR: Do you have him under control?

FIRST OFFICER FROM REAR: I'm okay.

TOWER: Express 705 heavy, can you proceed, direct to Memphis. Descend at your discretion and, uh, the localizer is on for runway niner.

CAPTAIN TO TOWER: All right, we're headed that way now, I think.

TOWER: Roger. Express 705 heavy, is the situation under control or is it still in progress?

CAPTAIN TO TOWER: We appear to have it under control.

TOWER: Roger.

CAPTAIN TO TOWER: Uh, Memphis, this is 705, understand that we need some medical personnel to meet us also!

TOWER TO CAPTAIN: Express 705 heavy, roger, that's being taken care of. They'll meet you there. Express 705 heavy, are you able to turn towards the airport?

CAPTAIN TO TOWER: Yeah, give me a vector.

TOWER TO CAPTAIN: Zero vector . . .

CAPTAIN TO TOWER: We're turning towards the airport now . . .

TOWER TO CAPTAIN: Left turn heading 100.

CAPTAIN TO TOWER: 100 Express 705. Express

705, I got to descend down to 7000 to proceed to Memphis.

TOWER TO CAPTAIN: Express 705, roger, descend at your discretion. Express 705 heavy, if able you can pick up the localizer runway niner and track it inbound.

CAPTAIN TO TOWER: Give me that frequency, please.

TOWER TO CAPTAIN: Okay, runway niner localizer is, uh, 109er.5.

CAPTAIN TO TOWER: 109.5, thank you, nine six, 89 inbound?

TOWER TO CAPTAIN: Say again?

CAPTAIN TO TOWER: 089 inbound?

TOWER TO CAPTAIN: Affirmative.

FIRST OFFICER FROM REAR: Dave!

CAPTAIN TO REAR: Yeah!

FIRST OFFICER FROM REAR: Can you get her [the aeroplane] on the ground?

CAPTAIN TO REAR: Okay.

TOWER TO CAPTAIN: Express 705 heavy, when you can I'd, uh, like to know your fuel on board and, uh, number of, uh, persons on board.

CAPTAIN TO TOWER: Okay, we'll, uh, give it to you in just a second.

TOWER TO CAPTAIN: No rush.

CAPTAIN TO REAR OF COCKPIT: Listen, is he okay, put, put that thing in his throat, I don't give a shit if he's dead or not, don't kill him but hold him, you got him, Jim? Jim, are you under control? Jim, are you under control? Jim? Jim, are you under control? Are you under control?

FIRST OFFICER FROM REAR: No, no! . . .
Something's the matter with me!
CAPTAIN TO REAR: There is something wrong with
him!
FIRST OFFICER FROM REAR: No!
CAPTAIN TO REAR: You keep him down, hear!
FIRST OFFICER FROM REAR: I can't . . . !
CAPTAIN TO REAR: You can keep him down, put
that thing in his . . . !
FIRST OFFICER FROM REAR: No, no!
TOWER TO CAPTAIN: Express 705 heavy, is that
localizer coming in now?
CAPTAIN TO TOWER: Yeah, we're on the localizer
now, descending.
TOWER TO CAPTAIN: Roger, and you want a visual
or do you just want to, do you want to shoot the ILS
[instrument landing system] or just shoot a visual?
CAPTAIN TO TOWER: I'll follow the ILS down and
take a visual.
TOWER TO CAPTAIN: Roger, Flight 705 heavy, at
pilot's discretion maintain, uh, 2000 and advise when
you get the airport in sight.
CAPTAIN TO TOWER: Will advise.
TOWER TO CAPTAIN: Okay, you're three zero, three
one miles west of the airport.
CAPTAIN TO TOWER: Thank you, sir.
CAPTAIN TO REAR: You got him down okay? Hey,
you put that, you keep him under control. Is he trying to
get up? [Sound of hammers being thrown into cockpit]
FIRST OFFICER FROM REAR: No!
CAPTAIN TO REAR: You hang in there now! Hey, put

that, go back and, hit him on the head, just . . . [Sounds of struggle from rear of plane]

TOWER TO CAPTAIN: Express 705 heavy, you're about twenty-five miles from the airport, and I'll be making a transmission every thirty to forty seconds just to stay in touch. [Sounds of struggle from rear of plane]

FLIGHT ENGINEER FROM REAR: Stay down!

CAPTAIN TO REAR: If you have to, if you have to put that in his throat, you do it! [Sounds of struggle from rear of plane]

CAPTAIN TO REAR: Is he still down?

FIRST OFFICER FROM REAR: Yeah, yeah!

CAPTAIN TO REAR: Is he, is he under control?

FIRST OFFICER FROM REAR: I don't know . . . Yeah, he is.

TOWER TO CAPTAIN: Express 705 heavy, you're twenty miles from the airport and uh, do you have that fuel and passenger information?

CAPTAIN TO TOWER: I got four on board, 86, uh, 85,000, I think 86,000 in fuel, four souls.

TOWER TO CAPTAIN: Roger Express 705 heavy, how many people should security be looking for?

CAPTAIN TO TOWER: Four.

TOWER TO CAPTAIN: Yeah, I mean how many involved in the action?

CAPTAIN TO TOWER: Everybody's been injured, uh, there's one person that, uh, lost it the, uh, jump-seat passenger's the one that attacked the crew.

TOWER TO CAPTAIN: Okay, thank you. [Sounds of struggle from rear of plane]

TOWER TO CAPTAIN: Express 705 heavy, verify uh, situation's still under control.

CAPTAIN TO TOWER: Well, it's sort of under control.

TOWER TO CAPTAIN: Okay, Express 705 heavy, uh, fifteen miles from the airport, about fourteen miles, uh, advise when you get it in sight.

CAPTAIN TO TOWER: I have it in sight.

TOWER TO CAPTAIN: Express 705 heavy is cleared visual runway 9, Express 705 heavy, the wind is uh, zero three zero at five, cleared to land runway niner. [Sounds of struggle from rear of cockpit]

CAPTAIN TO TOWER: Cleared to land?

CAPTAIN TO REAR: Kill the son of a bitch! Kill him! Kill him! Kill him! Kill him! [Sounds of groans from first officer in rear of cockpit]

FLIGHT ENGINEER FROM REAR: Jim, Jim, Jim, [sounds of struggle], Jim, help me! [Sounds of struggle, groans from first officer]

TOWER TO CAPTAIN: Express 705 heavy, you're about six and a half miles from the threshold, if able, when you get it on the ground, advise when you're on the ground, uh, I won't, uh, make any more transmissions to you at this time. [Sounds of struggle from rear of cockpit]

FLIGHT ENGINEER FROM REAR: Stop fighting! [The captain knows that the fight behind him in the cockpit is not over, and considers putting the aeroplane on autopilot at 7000 feet while he puts an end to the struggle. Instead he chooses to land the aeroplane and switches to runway 36L.]

CAPTAIN TO TOWER: I'm coming around to 36 left.

TOWER TO CAPTAIN: Okay, Express 705 heavy, runway 36 left, cleared to land, cleared visual [to runway] 36 left. You are cleared to land, the wind is 050 at 8. [Sounds of struggle from rear of cockpit]

FLIGHT ENGINEER FROM REAR: Ow! Jim, he's biting me!

FIRST OFFICER FROM REAR: Stay down!

[Groans from first officer in rear] [Sounds of struggle in rear of cockpit]

AUTO WARNING HORN: *Bank angle, bank angle.*

[Groans from first officer in rear]

TOWER TO CAPTAIN: Express 705 heavy all of the emergency equipment will be on frequency 121.9.

CAPTAIN TO TOWER: [Two clicks on microphone] [Sound of struggle in rear]

AUTO WARNING HORN: *Bank angle, bank angle.* [Sounds of struggle in rear]

FLIGHT ENGINEER FROM REAR: He's after the hammer, Jim!

AUTO WARNING HORN: *Altitude alert: one thousand.*

UNIDENTIFIED FROM REAR: Where's he going?

AUTO WARNING HORN: *Bank angle, bank angle. Too low! Terrain, sink rate, pull up, too low, terrain, sink rate 500, too low, terrain, sink rate* [sounds of struggle in rear]. *Pull up! Sink rate, pull up, sink rate, pull up, sink rate, sink rate . . .*

CAPTAIN TO TOWER: Get the crews over here now, get 'em over here in a hurry!

FLIGHT ENGINEER FROM REAR: Stop the jet, help us, stop the jet on the ground and help us!

TOWER TO CAPTAIN: Express 705, uh, help is on the way and frequency change approved, uh, the emergency equipment's on, uh, 121.9.

FLIGHT ENGINEER: Have they got the equipment out here?

CAPTAIN: They're on the way . . .

FLIGHT ENGINEER: Blow the door!

CAPTAIN: Yeah! [Sound of door being opened]

FLIGHT ENGINEER: Don't get close enough that he can grab anything!

CAPTAIN: Help me out . . . Don't move! Don't even think about it!

FLIGHT ENGINEER: Shut the engine down! Did you shut the engine down?

CAPTAIN: Yeah. Don't let him move!

FIRST OFFICER: Don't move!

[Sound of engine shutting down]

Calloway was convicted on a two-count indictment of air piracy and interference with flight operations and sentenced to life imprisonment with no possibility of parole. He is currently residing at the federal penitentiary in Atlanta.

The flight crew survived the attack. The captain suffered multiple lacerations to his head; he had been stabbed in his right arm and had a dislocated jaw. His right ear had been almost completely severed. The first officer's skull was fractured; he recovered from the paralysis to his right side but would experience ongoing motor-function impairment to his right arm and leg. He was blinded in one eye. The flight engineer suffered a skull fracture, as well as a severed temporal artery. The crew would never fly again.

CHARLOTTE, North Carolina, USA

2 July 1994

With fifty-two passengers, two cockpit crew and three flight attendants aboard, US Air Flight 1016 was executing a 'missed approach' while attempting to land on runway 18R at 6.42 p.m. in 'meteorological conditions'. It was raining hard. 'Convective activity that was conducive to a microburst [wind shear]', according to the NTSB, punched the Douglas DC-9-31 to the ground, where it collided with trees, breaking up, skidding like a toboggan down a residential lane and smashed into a private residence, catching fire. The captain and one flight attendant suffered minor injuries. The first officer, two flight attendants and fifteen passengers sustained serious injuries. The remaining thirty-seven passengers died. Impact forces and a post-crash fire destroyed the aeroplane. No one on the ground was injured or killed.

Despite a thunderstorm 'cell' advancing on the area, the cockpit crew had every reason to believe that a smooth approach and landing were possible.

COCKPIT VOICE RECORDER

APPROACH: Tell you what, USAir 1016, [you] may get some rain just south of the field. Might be a little bit just coming off the north. Just expect the ILS [instrument landing system] now. Amend your altitude . . . maintain three thousand.

CAPTAIN: If we have to bail out [go around for a missed approach], it looks like we bail out to the right.

FIRST OFFICER: Amen.

CAPTAIN: Ten miles to the VOR [navigational aid] which is off the end of the runway. 'Bout a mile off the end of the runway.

FIRST OFFICER: Yeah.

CAPTAIN: So I think we'll be all right.

CAPTAIN: Chance of [wind] shear.

FEMALE PASSENGER, AGED TWENTY-EIGHT, WHO WAS TRAVELLING WITH HER NINE-MONTH-OLD DAUGHTER, SITTING IN SEAT 19-F

During the flight my daughter moved back and forth between her seat and my lap. She wanted to play with the people in the row behind her. She got tired and laid down next to me. Her head was in my lap. After the flight attendant collected the lemonade cups, I felt a little bump and then a big bump, and the aeroplane just dropped. I could not understand what happened. The weather had been sunny, and I had seen thick white clouds. I heard an announcement, 'I'll have us on the ground in about ten minutes', and, 'Flight attendants, please prepare for landing'. I recall entering rainy weather, and I leaned forward in my seat to look out the window. The rain was coming on the wing slanted.

Other traffic in the sky near the airport and on the field's taxiways was aware of the storm, talking about it with the ground controllers and the tower.

COCKPIT VOICE RECORDER

TOWER: And [USAir] 806, looks like we've gotten a storm right on top of the field here.
US806: [On the ground waiting to depart] USAir 806, affirmative. We'll just delay for a while.
TOWER: [TO USAir 1016]: Charlotte Tower, runway 18 right, cleared to land. Following FK [Fokker] 100 on short final. Previous arrival reported a smooth ride all the way down final.
US 1016: USAir 1016, I'd appreciate a pirep [pilot flight report] from the guy in front of us.
FIRST OFFICER: Yep, [the storm is] laying right there this side of the airport, isn't it?
CAPTAIN: Well.
FIRST OFFICER: The edge of the rain is, I'd say . . .
CAPTAIN: Yeah.
TOWER: USAir 1016, company FK100 just exited the runway, sir; he said smooth ride.
TOWER: USAir 1016, wind is showing 100 at 19.
FIRST OFFICER: One hundred at 19, eh?
TOWER: USAir 1016, wind now 110 at 21.

CAPTAIN: Stay heads up.

TOWER: Wind-shear alert, north-east boundary winds 190 at 13. Carolina 5211, Charlotte Tower, runway 18R, cleared to land, wind 100 at 20. Wind-shear alert, north-east boundary wind 190 at 17. USAir 806, you want to just sit tight for a minute, sir?

US806: Yes, sir, we'd like to just sit tight.

TOWER: USAir 797, company aircraft in front of you is going to sit and wait awhile, sir. Do you want to go in front of him?

US797: No, no, it wouldn't sound like a good plan. We'll, uh . . . It didn't look like a whole lot [of rain] to us on the radar taxiing out, so it shouldn't be, uh, shouldn't be too many minutes.

CAPTAIN: [USAir 1016]: Here comes the wipers.

FIRST OFFICER: All right.

[Start of sounds similar to rain and sounds similar to windshield wipers on CVR]

FEMALE PASSENGER, AGED FORTY-FOUR, SEAT 19-D

I felt the engines, brakes or whatever, and then I felt a spinning like turning the engines to go, like [the captain] was circling. I was sitting next to the engine. I heard slowing down of the engine, and then the gunning up on the engine.

FEMALE PASSENGER, AGED TWENTY (WITH A NINE-MONTH-OLD INFANT), SEAT 21-C

The flight attendant kept peeking around the corner and smiling and making my baby laugh. I was seated in an aisle seat in the last row in front of the flight attendant. There was a little boy on the left side of the aeroplane at the window in front of me, and no one else was to my right. We were the only occupants of the row.

I heard the pilot make an announcement for landing and say that it was 'ninety-something' degrees and sunny out.

COCKPIT VOICE RECORDER

FIRST OFFICER: There's, oh, ten knots right there.
CAPTAIN: Okay, you're plus twenty. Take it around; go to the right.

At this point, the captain orders the first officer to abort the landing. On a missed approach they choose to go around to try to land again and, as such, they are 'on the go'. The aircraft is about 200 feet off the ground.

US1016: USAir 1016's on the go.
CAPTAIN: Max power.
FIRST OFFICER: Yeah, max power . . .
TOWER: USA 1016, understand you're on the go, sir. Fly runway heading. Climb and maintain three thousand.

FEMALE PASSENGER, AGED TWENTY-EIGHT (WITH HER NINE-MONTH-OLD DAUGHTER), SEAT 19-F

My seat belt was pulled so that it was fitting me. When I realized something was wrong, I leaned forward and pulled my daughter towards me. [She demonstrated to an NTSB Survival Factors Group interviewer a motion of leaning forward and wrapping her left arm over the child.] I thought I could use both arms to grasp my daughter as I leaned over.

MALE PASSENGER, AGED TWENTY, SEAT 21-D

Then suddenly we were in the middle of a dark rain cloud. As soon as we entered the dark cloud, the pilot goosed the gas and sped up. Then I felt the plane bouncing and then a jolt. In a split second everything opened up.

COCKPIT VOICE RECORDER

CAPTAIN: Down, push it down.
US1016: Up to three, we're takin' a right turn here.
TOWER: USAir 1016, understand you're turning right?

The flight's altitude begins decreasing below 350 feet. A second later, a 'whoop, whoop, whoop, terrain' warning sound begins and continues until the first sound of impact.

CAPTAIN: Power.

[Sound of impact]

FLIGHT ATTENDANT RICHARD DeMARY

(A copy of DeMary's interview was published in the March 1995 issue of *Cabin Crew Safety*, a publication of the Flight Safety Foundation.)

Probably the first real sign of trouble . . . was the sinking sensation, knowing that this is just not right . . . hearing – knowing we were off the airport – and hearing 'Terrain Terrain Terrain' [recorded warning from the cockpit].

Hearing that, and then [followed] almost immediately [by] the impact and fortunately or unfortunately, you know, being conscious through the whole thing, the whole crashing process began. It happened so fast. Initially it was disbelief and then just the terrifying feeling that we're crashing.

My recollection was that there were two impacts. You know, some people say there were three, but I remember two. The first impact with the ground, the sound of trees breaking, at that point knowing we were crashing – just the force of the impact was extremely violent, almost takes your breath away when you're crashing like that. Then immediately after the first impact, the second [and] most violent impact, to me, is when I think we hit a tree. The aeroplane hit a tree, basically peeled back [that] one side of the aeroplane – broke the aeroplane apart into three sections. The nose section, with a few passenger seats, went off to the left.

I was in that section, the nose section. One part of the aeroplane – and I believe it was the centre part of the aeroplane – from the first class seats back to just past the emergency exit rows, I believe – I don't know if this is proper to say – but it basically wrapped around a tree because that's what happened. We hit a tree and [the aircraft broke into separate sections] and then the tail section proceeded to go into [the carport of a house].

FEMALE PASSENGER, AGED TWENTY-EIGHT (WITH HER NINE-MONTH-OLD DAUGHTER), SEAT 19-F

After the crash, I remember pulling on my baby's leg to get to her. I was in and out of consciousness and did not remember how I got out of the aeroplane. It was difficult for me to open my eyes fully. When I came to, I saw blood on my daughter's face. I had dreamlike memories of hearing a woman asking, 'Where is my baby? Where is my baby?'

FEMALE PASSENGER, AGED FORTY-FOUR, SEAT 19-D

I recall a violent shaking and a big burst of fire made me think that I was going to die. There may have been up to five jolts. I was thrown up, back, over, back, and then the aeroplane came to a stop. Immediately after the impacts, flames blew up into my face and singed my hair and the parts of my body that were not covered by my clothing. There was a fire in the aeroplane when it came to a dead stop. Then the fire receded.

I looked around and saw that I was the only one in that part of the cabin. It was empty, dark and totally quiet. Then I began to hear people saying, 'Please help me. Please help me. Please help me out. I can't find my baby. I can't find my baby.'

FLIGHT ATTENDANT RICHARD DeMARY

I remember just after we hit the tree, the feeling of the rain hitting us, the wind, the noise, because we were in a darkened, enclosed cabin as we were coming in to land and then all of a sudden we were opened up to all of the elements and I remember feeling the rain hit me and the noise – screeching of metal, hearing the rain hitting the aeroplane, feeling it hit me – and it happened so fast that I guess I've never really had the opportunity to sit down and think about all the different sounds, but just noise. Probably the loudness came from the scratching of the aeroplane on the ground and on the pavement because the section I was in slid down a street.

I don't remember hearing people. I don't remember that now. Whether or not I ever will, I don't know.

I don't remember the wind, but I do remember the rain and I remember having jet fuel on me. Where it came from I don't know, maybe from puddles of jet fuel around or a spray of jet fuel or something. I'm not sure. But, really, nothing came to mind immediately other than [that] I couldn't open my door and there was no reason to open my door. I was in the open.

There was no cabin there. *There was no more aeroplane.* There were a few rows of seats on the nose section of the aeroplane that I was in, but the aeroplane had actually broken apart and bent.

I remember . . . sitting in my jump seat, not seeing anything once the aeroplane came to a stop, and at that point knowing we were in a crash, knowing that it's time to get out, it's time to evacuate, and I immediately went for my seat belt [and] started yelling, 'Release seat belts and get out! Release seat belts and get out!' which is the first command that we would yell upon coming to a stop, and at that point Shelly [Markwith, a flight attendant sitting in the jump seat beside DeMary] . . . was yelling her commands, 'Release seat belts and get out!' and trying to get her seat belt unbuckled. Shelly told me, 'I can't get out. My legs are broken. I can't get out.'

I was actually leaning on Shelly. I stood up and I had to kick my feet free from the debris. As I stood up, I believe I saw the captain crawl out of the cockpit, through the cockpit door. Because of Shelly's injury she couldn't do anything. She had shattered her kneecap and I think she had a cut to her bone on her thigh, and she had some burns and she – she couldn't even crawl.

Shelly was having some difficulty getting herself unbuckled, so I unbuckled her seat belt, bear-hugged her, grabbed her and picked her up. She couldn't stand because her leg was severely injured and she really couldn't do anything to evacuate herself. So I grabbed her and just carried her and stepped off to the street. Five feet away from the aeroplane she fell again, and then I grabbed her wrists and just dragged her away, just trying to get her to a safe distance away from the aeroplane.

FEMALE PASSENGER, AGED TWENTY (WITH NINE-MONTH-OLD INFANT), SEAT 21-C

Then there was a bump, and my baby was crying as she flew out of my arms. I tried to hold onto her. There was a bump. There were a lot of bumps. The wheels touched the ground, and we were bouncing all over the place, and I hit my head on something. People were flying all over the place.

MALE PASSENGER, AGED TWENTY, SEAT 21-D

I felt stuff, dirt, hitting me in the face. The next thing I remembered was laying on my back under a pile of metal with everything piled on top of me. I barely had enough room on my chest to exhale and inhale. My lung had collapsed.

I felt like I was in the wreckage for about an hour before I was pulled out. I was on my back, facing upward. I could only see metal and a little bit of light. I felt a lot of heat on my face. I could hear a few people screaming and asking for help. After a few minutes of lying in the wreckage I heard someone say, 'Is anyone in there?' Four or five people yelled out, and I yelled out, too. It was so tight in there that I really couldn't yell that loud because I couldn't get enough air.

Then the awareness of being in the accident, that now I have survived, that I have to do something, came full force. It became very important for me to help anybody I could help, not only to help but to search, to find people, and I couldn't wait for them to come to me. I had to go find somebody. I was completely disoriented as far as where the rest of the aeroplane was. At that point, the thought crossed my mind that we were the only ones who survived.

[The] fires were bad enough that I could feel the heat, so I knew that we had to get away or the people [who] were too injured to do anything [to escape] had to get away from [the heat]. It was – it was hot! A lot of fires were spotted around the area, a lot of small fires, and then I remember the smoke and feeling the flames and seeing the flames over quite a large area which turned out . . . to be by the tail cone, the back of the aeroplane. There was a lot of fire. [The nose section] broke off and came to a stop in the street just to the left of a house – in front of the house. I helped Shelly away into a grassy little area of yard and I was confident that she was safe at that point. But after helping Shelly . . . I had no aeroplane. There was just nothing there.

I [suddenly realized] that, 'Oh, my God, this is a residential neighbourhood!' Because I saw the houses. I saw the trees. I saw the street, the sidewalk, and I think I immediately thought, 'What are we doing here? This is not right!' Because I always thought, 'Well, it's reality that we might crash sometime', but I never thought it would be into a house or into a residential neighbourhood. I mean, we were in *somebody's yard*.

FEMALE PASSENGER, AGED TWENTY-EIGHT (WITH HER NINE-MONTH-OLD DAUGHTER), SEAT 19-F

I awoke outside the aeroplane where someone had dragged me away from the wreckage. When I awoke my head was in the lap of a man . . . My daughter was being held by a woman. I heard my daughter crying, and I told her, 'Mom's here.'

FEMALE PASSENGER, AGED FORTY-FOUR, SEAT 19-D

I believe that instinct kicked in when I realized there was smoke. I tried not to breathe any smoke by taking only shallow breaths. I told myself, 'Don't panic! You can get out!' I kept trying to focus on, and believe in, my ability to get out. I believed that my positive thinking helped me to survive. I did not panic. I did not want to die, and I intended to do everything necessary to prevent it. I decided that if I was going to die, then it was God's will.

I unfastened my seat belt and stood up. It's too incredible to explain the position of all the seats. The seats were down under [upside down?]. I stood up and looked around to position myself. From my back right side I saw the flight attendant crawl over a bunch of stuff. I asked the flight attendant, 'What can I do to help?' She replied, 'Come and try to help me open this back rear emergency door.'

About the same time a black man who had been sitting to my left escaped from under his seat. He crawled up through the wreckage and also came back to help the flight attendant.

FLIGHT ATTENDANT RICHARD DeMARY

At that point I took off my tie. I don't have a memory of little bits of what I did. It's my understanding there was a lot of fire and possibly a lot of bodies, you know, my mind just doesn't want me to have it right now. But I remember ending up by the tail section and it was very quiet. I didn't hear anybody, didn't see anybody. There was a break in the aeroplane, a break in the fuselage, and at that point, I thought, 'Well, I have to do something!' and I started yelling my commands. I thought, 'Well, it's a starting point. If [people are] in shock, if they hear, "Release seat belts and get out!" it's going to give them the starting point.' So I started yelling, 'Release seat belts and get out! Release seat belts and get out!' I'm continuously yelling it, as I'm walking, as I'm looking for somebody, looking . . .

I didn't actually go [back] in the aircraft. I was right beside it, right next to the engine. There was just a small break in the right side of the fuselage. That side was fairly intact.

I had really given up at one point. I thought, 'Well, there's probably nobody that survived – that survived the impact', but I remained. I continued to have faith that somebody might have survived – you know, somebody might be in there. I remember how hot it was. The fire was tremendously hot.

Then a woman appeared at that break [in the fuselage] with a baby. She was able to get out of her seat belt and this was probably some time after the accident [before any rescue squads had arrived on the scene] . . . It seemed like an eternity but she came towards my voice.

. . . She appeared at that small opening and I reached in and grabbed the baby . . . and grabbed her arm and pulled her. I mean, it wasn't like Shelly. I literally [had] dragged Shelly on the ground, but I'm sure I just grabbed [this woman with the baby]. She was yelling, 'Help me! Help me!' when she was in the aeroplane.

There was a small shed in the back yard [behind the house carport, which the piece of aircraft fuselage had struck] and I just took them back there to safety.

I went back to the aeroplane again, and I remember thinking how hot it was – I mean, I placed my arm on the engine [cowling] and it just burned all the skin off my – not the skin but it severely burned my arm and just the heat of it, the heat of the metal, and I remember hearing explosions, small explosions, and I thought, 'Well, I have to do what I have to do, but I can't stay here forever.' I was concerned that I would succumb to the smoke or, you know, the fire or something like that.

But anyway, I did go back, and I continued to yell, 'Release seat belts and get out!' and another woman appeared at the same opening . . . She was yelling, 'I don't want to die! Help me! I don't want to die! I can't find my baby!'

I later learned that she was one of the women who had a child, a lap child, with her. [Of the two lap children on the flight, only one survived.] I helped her out of the aeroplane and she had some injuries, I think, because she was basically immobile. It took a lot to get her out. The tail of the aeroplane was on the ground but the centre section was in the air. It was at quite an angle. So the opening was probably . . . mid-waist to chest-high.

I had to reach up just a little. She's yelling, 'I don't want to die! I don't want to die! Help me out! I can't find my baby! I can't find my baby!' I literally had to just bear-hug her and pull her out because she was heavy. Anyway, I got her out and got her back to the same place I took the others at the back of the yard.

You know, she was yelling, 'I can't find my baby!' and I went back then, but at that point I didn't think it was probably . . . [the] best thing to do. What I'm saying is, I didn't think it was appropriate that I was actually going into the aeroplane and search[ing], because of the fire and the smoke and how long it took them to get out.

But I [started] back again after helping the lady out and [I saw a man and a woman] from the neighbourhood. I asked them just to stay with the passengers [in the back yard] and I went back to the aeroplane and continued to yell . . . There was nobody. And at that point, well . . . I thought, 'I need to get away because it's very hot and I don't want to survive the impact to die in the fire of the secondary explosions.' Something like that. And I thought I could be of help somewhere else, possibly.

And one of the things that bothered me, too, is that I did have jet fuel on me. My clothing was flammable and probably more so with jet fuel on me. So I went back around the back side of the house, towards the front yard, and I saw the captain, and at that point Fire and Rescue still hadn't arrived. I remember hearing the captain say, 'She's okay! She's okay!' I thought he was talking about Shelly but in reality he was talking about Karen [Forcht, the third flight attendant on the aeroplane], and then I did see Karen, and she had severe burns. She had lost her shoes in the impact and she had severe burns on her arms, hands, face, legs. I believe three people had followed Karen out when she got out.

FEMALE PASSENGER, AGED FORTY-FOUR, SEAT 19-D

We pulled the door open a little, and flames were visible. The flight attendant said, 'No. No.' And we shut the door immediately. Then the flight attendant turned around and said, 'We can't get out this way.' And we went the other way.

By the time I refocused and turned around there was nobody else there. I saw the [same] black man and a small black child wiggling out of a place where there was light at the tail end of the aeroplane. I had previously seen this area of light when I went to help with the door, but I ignored it because I had seen flames as well as light. *[Just as Flight Attendant DeMary had trouble believing that the aeroplane had come to a halt in a neighbourhood with trees and houses, this passenger was equally baffled, and reasonably so, when she looked down before stepping out of the wreckage and saw a household's kitchen in front of her.]* I came back to where my seat was, and I noticed that to the left there was a door going into the kitchen [of the house that the aircraft had collided with]. I was confused.

55

I asked myself, 'Why is this kitchen here?' I tried to open the kitchen door, thinking that I could get out through the kitchen. The door was like a storm door with glass on the top and a white bottom. The glass in the door was not broken. I wanted to smash the window, but I could not find anything loose that I could use.

I decided that I [had to escape] through the area that I had seen the man and his young son wiggling through. Then I heard a man yell, 'We made it! We made it!' And I knew I had to go that way.

FEMALE PASSENGER, AGED TWENTY [WITH A NINE-MONTH-OLD INFANT], SEAT 21-C

During the crash and the impact of the aeroplane hitting the ground, my baby went flying in front of me. I tried to hold her and I couldn't. They had told me I could hold her in my lap. I would have paid for her to sit in a seat. They said she did not need a seat. The man said that she did not need a seat because she was under the age of two, and that she was a 'lap baby', and I could hold her. I would have given my life for her. She wasn't here that long. She had just turned nine months old.

MALE PASSENGER, AGED TWENTY, SEAT 21-D

I could hear people above me. It sounded like people [were] talking on metal and banging metal around. I heard someone say, 'Where are you?' And all I could say was, 'Here!' because I did not know where I was. Someone put an air mask on top of my face. I could hear cutting, and when they used the cutters I could feel vibrating around me. They finally got to me and cut me out.

FEMALE PASSENGER, AGED FORTY-FOUR, SEAT 19-D

My seatmate [in 19-F], whom I had not seen before, now came to and began yelling, 'Please help me, please, somebody help me.' He had a tree lying on top of him, and I could only see his head and one boot. I asked him, 'What do you want me to do?' I could not get him out. I told him to try to think of something else. I knew I could get out, and I would send some help.

I did not think the [fireball] caused my burns. I think that I got my burns when I shimmied out of the aeroplane. Everything that was metal was so hot it was like touching a hot griddle. As I got out through the tail I was still about eight feet off the ground. I saw the hood of a car that was slanted, and I slid off of it and ended up at the front porch of the house that was connected to the carport.

FEMALE PASSENGER, AGED TWENTY [WITH A NINE-MONTH-OLD INFANT, WHO HAD DIED], SEAT 21-C

I heard my baby calling to me after the aeroplane came to a rest. She was calling me: 'Mama.' A man . . . pulled me out of the plane and told me that my baby was okay. They said she was okay and that she was at another hospital.

[The woman was unable to restrain her child during the impact and her child was thrown forward over three rows of seats and killed by trauma.]

FLIGHT ATTENDANT RICHARD DeMARY

At that point a few other passengers were coming out of the wreckage. Usually, when you think about an accident, you think everybody's going to be going through the same thing. Karen's [experience] of the accident was a little bit different because she was in the back of the aeroplane that broke apart. She had the impact. She had the debris flying through the cabin. She had the fireball. She had the smoke. So she had a lot of different elements to contend with. And then probably most importantly, Karen had the element of not having a usable exit. She was able with the assistance I think of a couple of passengers to get that back [emergency] door open [to] access the tail cone and found it unusable, [and] immediately closed [the door]. That's where the fire was. There was smoke in there.

There was a triage area forming, basically, to the left of the house. So there was no sense in me staying with the people [who] were, I guess, okay. I went back towards the front of the house and I remember seeing a kid from the neighbourhood – this thirteen- or fourteen-year-old kid – and I said, 'Is anybody at home? Is anybody in the house?' and he said he didn't know, so I thought, 'Okay, the next thing to do is to go into the house, because the aeroplane was too hot, there was too much fire and I wasn't going to go inside the aeroplane.'

I'm talking to that young kid in front of the house and

I thought, 'Well, if anybody's home, we need to see about them.' So I went to the front door and just as I was running to the front door, a passenger crawled out of the wreckage who said, 'Somebody's in the garage.'

I thought, 'Well, there are people home. There is somebody there who needs help!' There were cars within the wreckage, from the driveway, so I went to open up the front door and I thought, because it was locked, there's probably nobody home. So I kicked in the front door and just looked to the left a little bit and the captain followed me in and I think the young boy from the neighbourhood followed me in. There didn't appear to be any damage to the left inside of the house.

Then I looked right and I saw the living area or the dining room area with a set table. I remember seeing the [place] mats on the table, and then I looked and I saw the door that [opened onto] the garage. It was actually just a carport, and it opened to the inside and then there was a [storm] door, and [the door] opened to the outside. I couldn't open up that door because of the debris within the garage area and – so then I just busted out the glass of the door.

And then I heard a voice [in the carport], and then I started just speaking with the guy, yelling to him. He was yelling, 'Help me! I can't breathe.' And I was yelling back to him, 'Cover your mouth if you have anything to cover your mouth with, and breathe through that', and he was yelling back, 'I don't have anything to cover my mouth with.' I couldn't see him because the smoke was very heavy.

It was kind of – it was a greyish smell of, like, plastic burning, just very heavy. I couldn't even breathe it in, and

I was getting good air [from inside the house] with that bad air [in the carport], and I found it difficult to breathe.

There was so much debris, and I remember one of the main-wheel tyres was standing right there next to the house, just within the debris, and I couldn't see the guy. I couldn't really make out anything in that area, and I yelled for him to cover his mouth if he had anything to cover his mouth with, and he said he didn't, and then I shouted for him to stay calm, try to relax, breathe slowly, just to stay calm, that help was on its way. At that point I heard the fire trucks arriving. I said [to the man], 'Somebody's here to help.'

I ran to tell [the rescuers] that somebody was in that carport area. The fire trucks couldn't get in because we did crash in a residential neighbourhood, and [the aircraft had] sheared some telephone poles. [Another guy and I] moved the telephone poles so that [the fire trucks] could get out, or get back in to a closer area, and I told them that there was somebody [in the carport]. The gentleman did survive. He was a passenger.

So the fire and the rescue trucks were arriving. [I] helped [the fire fighters] pull out some hose [from the fire trucks] and then was asked to get away. I was told, well, you know, 'Your job's done. Just get away.'

PITTSBURGH, Pennsylvania, USA

8 September 1994

USAir Flight 427, a Boeing 737–3B7, departed Chicago-O'Hare Airport for Pittsburgh at about 6.10 p.m. with two pilots, three flight attendants and 127 passengers. Flight 427 was approaching Pittsburgh for a landing on runway 28R when Pittsburgh air traffic control reported traffic in the area that Flight 427's first officer confirmed by sight. The aircraft was levelling of at 6000 feet at a speed of 190 knots and rolling out of a 15 degree left turn with flaps at 1, with its gear still retracted and autopilot and auto throttle systems engaged, when the aircraft suddenly entered the wake vortex of a Delta Airlines Boeing 727 that preceded it by approximately 69 seconds and 4.2 miles. Over the next three seconds Flight 427 rolled left to an approximately 18 degree of bank. The autopilot initiated a roll back to the right as the aircraft went in and out of a wake vortex core, resulting in two loud 'thumps'. But the first officer manually overrode the autopilot without disengaging it by making a large right-wheel turn. The aeroplane started rolling back to the right, but it never reached a wings-level attitude.

At 19:03:01 the aircraft slewed suddenly and dramatically to the left and pitched down and continued to roll through 55-degree left bank.

At 19:03:07 the pitch approached -20 degrees. The left bank increased to 70 degrees and the descent rate reached 3600 feet per minute. The aircraft stalled. Left roll and yaw continued, and the aircraft rolled through inverted flight as the nose reached 90 degrees down, approximately 3600 feet above the ground. The aircraft continued to roll, but the nose began to rise. At 2000 feet above the ground the aircraft's attitude passed 40 degrees nose low and 15 degrees left bank. The left roll hesitated briefly, but continued, and the nose again dropped. The plane descended fast and impacted the ground nose first at 261 knots in an 80-degree nose down, 60-degree left bank attitude and with significant sideslip.

18:32:29 CONTROL: . . . Contact Chicago's Cleveland Centre one two six point niner seven.
18:32:33 CAPTAIN: Twenty-six ninety-seven, USAir 427, good day.
18:33:08 CAPTAIN: Centre USAir 427 at two nine oh.
18:33:14 CLEVELAND CONTROL: USAir four twenty-seven, Cleveland Centre, roger.
18:33:32 CLEVELAND CONTROL: USAir 427, cleared direct to, uh, Akron, rest of route unchanged, give me the best forward airspeed in-trail spacing.
18:33:37 FIRST OFFICER: Direct Akron, best forward, you got it, 427 USAir.
18.37:46 [Sound similar to person yawning]
18:38:00 CAPTAIN: Blocked.
18:43:27 CLEVELAND CONTROL: USAir 427, contact Cleveland Centre one one niner point eight seven

nineteen eighty-seven, USAir four-twenty seven, good day.

18:43:37 CAPTAIN: Do you wanta let 'em [passengers] up for a while?

18:43:42 [Sound of single chime similar to seat belt switch being moved]

18:45:31 CLEVELAND CONTROL: USAir 427, descend and maintain flight level two four zero [24,000 feet].

18:45:35 CAPTAIN: Out of two nine oh for two four oh, USAir 427.

18:45:58 FIRST OFFICER: Ah, you piece of shit. What? I said, aw c'mon, you piece of shit. This damn thing is so damned slow.

18:46:07 FIRST OFFICER: There it is.

18;47:23 CLEVELAND CONTROL: USAir 427 contact Cleveland Centre one two eight point one five.

18:47:27 CAPTAIN: Twenty-eight fifteen, USAir 427. Good day.

18:47:35 CAPTAIN: Centre, USAir 427 descending two four oh [24,000].

18:50:18 [Aural tone similar to attitude alert]

18:51:22 PITTSBURGH TERMINAL INFORMATION SERVICE, RECORDED: ... Pittsburgh tower arrival information Yankee two one five two Zulu weather. Two five thousand scattered. Visibility one five. Temperature seven five. Dew point five one. Wind, two seven zero at one zero. Altimeter three zero one zero. Multiple approaches ILS runway three two and ILS runway two eight right in use. NOTAMS, runway 28 right middle marker decommissioned. South entrance Air Force Reserve ramp closed. Morgantown VORTAC out of service. Advise on initial contact you have information Yankee.

18:53:29 CAPTAIN: Three two and two eight right. Three two and two eight right?

Yep. [Sound similar to cockpit door being opened and closed] Um, they didn't give us connecting flight information or anything. Do you know what gate we're coming in on?

18:53:58 FIRST OFFICER: No. Do ya know what I'm thinkin' about? Pretzels.

18:56:36 CAPTAIN: USAir 427, reduce speed to two one zero now. That's at the request of Pit approach. I'll take the speed first. 427.

CAPTAIN: Uh, we'll do our best to make the restriction. Don't have to now. Just, uh, first uh, [reduce] speed to ten. You got it.

APPROACH: USAir 427, contact Pit approach one two one point two five.

CAPTAIN: Twenty-one twenty-five, USAir 427, good day.

18:56:49 CAPTAIN: I may have misunderstood him. Approach, USAir 427 is descending to ten.

18:56:55 APPROACH TO ANOTHER AIRCRAFT IN TRAFFIC: Shuttle twenty-nine zero eight, turn right heading zero niner zero. Contact approach one two three point niner five.

18:57:05 APPROACH: USAir 1429, turn left heading one zero zero. Contact approach one two four point one five.

18:57:08 [Sound similar to cockpit door being opened]
FLIGHT ATTENDANT: Here it is [drinks for the cockpit crew].
18:57:09 CAPTAIN: All right. All right. Thank you, thank you.
FLIGHT ATTENDANT: If you don't like, I didn't taste 'em so I don't know if they came out right [drink she had concocted for the cockpit crew].
18:57:14 APPROACH TO ANOTHER AIRCRAFT IN TRAFFIC: USAir 1874, descend and maintain six thousand.
18:57:14 CAPTAIN: That's good. That is good. It's good. That is different. Be real, be real good with some dark rum in it.
FLIGHT ATTENDANT: Yeah, right. Can I get you something else?
18:57:23 APPROACH: USAir 427, Pittsburgh approach. Heading one six zero vector ILS runway 28 right. Final approach course. Speed two one zero.
18:57:26 FIRST OFFICER: What kind of speed? Okay.
18:57:29 CAPTAIN: We're, comin' back to two ten and, uh, one sixty heading down to ten, USAir 427, and, uh, we have Yankee.
18:57:40 CAPTAIN TO APPROACH: What runway did he say?
18:57:43 APPROACH: USAir 1462, Pittsburgh approach heading zero four zero vector ILS three two final approach course.
18:57:49 FLIGHT ATTENDANT: It tastes . . . good. There's little grapefruit in it. [Sound of laughter]
18:57:54 APPROACH: USAir 1674, turn left heading

one zero zero. Contact approach one two four point one five.

18:57:55 FLIGHT ATTENDANT: . . . Cranberry. Yeah. You saw that from the colour.

18:58:03 FIRST OFFICER: What else is in it? Uh, Sprite?

18:58:08 FLIGHT ATTENDANT: And I guess you could do it with Sprite. Prob'ly be a little better if you do.

18:58:13 FIRST OFFICER: Yeah, there's more? One more. Okay?

18:58:14 APPROACH: USAir 374, turn right heading one one zero.

18:58:20 CAPTAIN: You got it. Huh. Cranberry orange and Diet Sprite. Really nice.

It's different. Could ya keep comin' out, aaah, whataya got different and I always mix the cranberry and the grapefruit, I like that.

18:58:26 FIRST OFFICER: [Sound of aural tone similar to altitude alert] Okay, back to work.

18:58:29 [Sound similar to cockpit door opening and closing]

18:58:29 FIRST OFFICER: I suspect we're gonna get the right side.

18:58:33 APPROACH: USAir 427, descend and maintain six thousand.

18:58:36 CAPTAIN: Cleared to six, USAir 427.

18:58:48 [Sound of click similar to approach plate clip being snapped]

18:58:50 FIRST OFFICER: Oh, my wife would like that [drink].

18:58:56 CAPTAIN: Cranberry orange and Sprite.

18:58:57 APPROACH: USAir 1674, contact approach one two four point one five.

18.59:06 CAPTAIN: Yeah. I guess we ought to do a preliminary, Pete. Altimeters and flight instruments thirty eleven?

18:59:30 APPROACH: USAir 374, contact approach one two.

18:59:41 APPROACH: USAir 309, Pittsburgh approach, heading zero five zero vector ILS runway three two final approach course.

18:59:54 CAPTAIN: Ah, don't do this to me.

18:59:56 FIRST OFFICER: [Sound of chuckle] Froze up, did it?

19:00:08 APPROACH: Delta 1083, turn left heading one three zero. Reduce speed to one niner zero.

19:00:12 CAPTAIN: I hate it when you don't hear the other transmissions.

19:00:14 FIRST OFFICER: [Chuckle] Yeah.

19:00:15 APPROACH: USAir 427, turn left heading one four zero, reduce speed to one niner zero.

19:00:20 CAPTAIN: Okay, one four zero heading and one ninety on the speed, USAir 427.

19:00:28 [Sound of three clicks similar to flap handle being moved] [Sound of single chime similar to seat belt chime]

FIRST OFFICER: Oops, I didn't kiss 'em 'bye. [Clicking sound similar to trim wheel turning at auto pilot trim speed]

19:00:31 FIRST OFFICER: What was the temperature . . . ?

19:00:34 CAPTAIN: Seventy-five. Seventy-five?

19.00:37 [Clicking sound similar to trim wheel turning at autopilot trim speed]

19:00:43 FLIGHT ATTENDANT: Seat belts and remain seated for the duration of the flight.

19:00:44 FIRST OFFICER TO PASSENGERS: Folks, from the flight deck we should be on the ground in 'bout ten more minutes. Uh, sunny skies, little hazy. Temperature, temperature's, ah, seventy-five degrees. Wind's out of the west around ten miles per hour. Certainly 'preciate you choosing USAir for your travel needs this evening. Hope you've enjoyed the flight. Hope you come back and travel with us again. This time we'd like to ask our flight attendants please prepare the cabin for arrival. Ask you to check the security of your seat belts. Thank you.

19:00:46 APPROACH: Delta 1083, turn left heading one zero zero.

19:00:48 DL1083 [DELTA AIRLINES 1083]: One zero zero, 1083.

19:01:04 CAPTAIN TO APPROACH: Did you say 28 left for USAir 427?

19:01:06 [Chime similar to seat belt chime]

19:01:06 APPROACH: Uh, USAir four twenty-seven, it'll be 28 right.

19:01:08 CAPTAIN: 28 right, thank you.

19:01:11 FIRST OFFICER: 28 right. Right, 28 right. That's what we planned on. Autobrakes on one for it.

19:01:18 APPROACH: Delta 1083 contact approach one two four point one five.

19:01:22 DL1083: Twenty-four fifteen, good day.

19:01:26 APPROACH: USAir 1462, at six thousand, reduce speed to one niner zero. USAir 309, descend and maintain six thousand then reduce speed to one niner zero.

19:01:42 FIRST OFFICER: Bravo thirty-nine . . . that's not too bad that's . . .

19:01:47 APPROACH: USAir 1874, turn right heading one zero zero. Contact approach one two three point niner five.

19:01:48 FIRST OFFICER TO APPROACH: 'bout halfway.

19:01:56 [Aural tone similar to altitude alert]
CAPTAIN: [Then] . . . [works] seven for six.

19:01:57 APPROACH: USAir 1462, turn right heading zero eight zero.

19:01:58 FIRST OFFICER TO APPROACH: Seven [thousand feet] for six. Boy, they always slow you up so bad here. That sun is gonna be just like it was takin' off in Cleveland yesterday, too. I'm just gonna close my eyes. [Sound of laughter] You holler when it looks like we're close. [Sound of laughter]

19:02:24 CAPTAIN: [Sound of chuckle] Okay.

19:02:27 APPROACH: USAir 427, turn left heading one zero zero. Traffic will be one to two o'clock, six miles, northbound Jetstream, climbing out of thirty-three for five thousand.

19:02:32.0 CAPTAIN TO APPROACH: We're looking for the traffic, turning to one zero zero, USAir 427.

19:02:32.9 [Sound similar to aircraft engines increasing in rpm to a steady value] [Clicking sound similar to trim wheel turning at autopilot trim speed]

19:02:54.3 FIRST OFFICER: Oh, ya, I see zuh Jetstream. [Sound similar to three electrical clicks within one second]

19:02:57.5 CAPTAIN: Sheeez.

19:02:57.6 FIRST OFFICER: Zuh.

19:02:58.1 [Sound of thump]

19:02:58.6 [Sound of 'clickety click']

19:02:58.7 CAPTAIN: [Sound similar to person inhaling/exhaling quickly once]

19:02:59.1 [Sound of thump of less magnitude than the first thump]

19:02:59.4 CAPTAIN: Whoa.

19:03:00.7 [Clicking sound similar to trim wheel turning at autopilot trim speed]

19:03:01.1 CAPTAIN: Hang on.

19:03:01.5 [Sound similar to aircraft engines increasing in rpm]

19:03:01.6 FIRST OFFICER: [Sound similar to pilot grunting]

19:03:02.0 CAPTAIN: Hang on.

19:03:02.1 [Sound of click and wailing horn similar to autopilot disconnect]

19:03:03.6 CAPTAIN: Hang on.

19:03:04.6 FIRST OFFICER: Oh, shit.

19:03:05.2 CAPTAIN: Hang on.

19:03:07.5 [Sound of increasing amplitude similar to onset of stall buffet]

7:03:08.0 CAPTAIN: What the hell is this?

19:03:08.1 [Vibrating sound similar to aircraft stick shaker starts and continues to end of recording]

19:03:08.3 [Sound of aural tone similar to attitude alert]

19:03:09.6 CAPTAIN: What the . . .
19:03:09.9 FIRST OFFICER: Oh . . .
19:03:10.6 CAPTAIN: Oh God . . . Oh God.
19:03:13.3 APPROACH: USAir . . .
19:03:15.0 CAPTAIN TO APPROACH: 427,
emergency.
19:03:16.1 CAPTAIN: Pull.
19:03:18.5 FIRST OFFICER: Oh, shit.
19:03:19.1 CAPTAIN: Pull.
19:03:19.7 CAPTAIN: [Pull]
19:03:20.8 FIRST OFFICER: God.
19:03:21.1 CAPTAIN: [Sound of screaming]
19:03:22.5 FIRST OFFICER: No.
19.03:22.8 END OF RECORDING

All 132 passengers and crew died as a result of the crash.

The NTSB determined that the probable cause of the USAir Flight 427 accident was a loss of control of the aeroplane 'resulting from the movement of the rudder surface to its blow-down limit. The rudder surface most likely deflected in a direction opposite to that commanded by the pilots as a result of a jam of the main rudder power control unit servo valve secondary slide to the servo valve housing offset from its neutral position and over travel of the primary slide.'

EIGHT MILES OFF EAST MORICHES, New York, USA
17 July 1996

TWA Flight 800, a Boeing 747 registered as N93119, was scheduled to depart New York/JFK for Paris at about 7.00 p.m. However, the flight was delayed because of a disabled piece of ground equipment and concerns about a suspected passenger/baggage mismatch. The aircraft was pushed back from the gate at about 8.02. Between 8.05 and 9.07 the flight crew started numbers 1, 2 and 4 engines and completed the after-start checklist. The flight crew then received taxi instructions and began to taxi to runway 22R. While the aeroplane was taxiing, at about 8.14, the flight crew started number 3 engine and conducted the delayed engine-start and taxi checklists.

At 20:18:21 air traffic control (ATC) advised the pilots that the wind was out of 240 degrees at 8 knots and cleared Flight 800 for takeoff. After takeoff the pilots received a series of altitude assignments and heading changes from New York Terminal Radar Approach Control and Boston ARTCC (Air Route Traffic Control Centre) controllers. At 20:25:41, Boston ARTCC advised the pilots to climb and maintain FL190 (flight level 19,000 feet).

At 20:26:24, Boston ARTCC amended TWA flight 800's altitude clearance, advising the pilots to maintain FL130. At 20:29:15, the captain stated, 'Look at that crazy fuel flow indicator there on number four ... see that?' One minute later Boston ARTCC advised them to climb and maintain FL150. The crew then selected climb thrust. After a very loud sound for a fraction of a second, the CVR stopped recording at 20:31:12. At that moment, the crew of an Eastwind Airlines Boeing 737 flying nearby reported seeing an explosion. The aircraft broke up and debris fell into the sea, eight miles south off East Moriches. The first transcript describes the conversation between Boston Centre (Control) and aircraft flying in the vicinity of TWA Flight 800.

BOS = Boston Centre
E507 = Eastwind Flight 507
AZA609 = Alitalia Flight 609
VIR009 = Virgin Atlantic Flight 009
UAL002 = United Air Lines Flight 2

20:31:50 E507: We just saw an explosion out here; stinger bee [Eastwind] 507.
20:31:57 BOS: Stinger bee 507, I'm sorry I missed it, ah, you're out of eighteen. Did you say something else?
20:32:01 E507: Ah, we just saw an explosion up ahead of us here, about sixteen thousand feet or something like that; it just went down – in the water.
20:32:10 AZA609: Alitalia 609, confirms just ahead of us.
20:32:25 VIR009: Boston, Virgin 009, I can confirm that out of my nine o'clock position, we just had an ... it looked like an explosion out there about five miles away, six miles away.
20:32:49 BOS: An explosion six miles out at your nine

o'clock position. Thank you very much, sir. Contact New York approach one two five point seven.

20:32:56 BOS: TWA 800, centre.

20:33:01 ? Investigate that explosion if you get a lat/long.

20:33:04 BOS: TWA 800, centre.

20:33:09 BOS: TWA 800, if you hear centre, ident.

20:33:17 BOS: Stinger bee 507, you reported an explosion is that correct, sir?

20:33:21 E507: Yes, sir, about five miles at my eleven o'clock here.

20:33:31 BOS: Alitalia 609, contact Boston now on one two four point five two.

20:33:36 AZA609: One two four point five two and just for your information, sir, we are just overhead the explosion; right overhead at this time, now a hundred and three miles from JFK.

20:33:48 E507: We are directly over the site with that aeroplane or whatever it was just exploded and went down into the water.

20:34:01 BOS: Roger that, thank you very much, sir. We're investigating that right now. TWA 800 if you hear centre, ident.

20:35:43 ? I think that was him.

20:35:45 BOS: I think so.

20:35:48 ? God bless him.

20:37:05 E507: Thirty-three oh five, so long, stinger 507. Anything we can do for ya before we go?

20:37:11 BOS: Well, I just wanna confirm that you saw the splash in the water approximately twenty south-west of Hampton, is that right?

20:37:20 E507: Ah, yes, sir. It just blew up in the air and then we saw two fireballs go down into the water and there was a big small, ah, smoke, ah, coming up from that also, ah, there seemed to be a light. I thought it was a landing light and it was coming right at us at about, I don't know, about fifteen thousand feet or something like that and I pushed on my landing lights, ah, you know, so I saw him and then it blew.

20:37:40 BOS: Roger that, sir, that was a seven forty-seven out there; you had a visual on that; anything else in the area when it happened?

20:37:47 E507: I didn't see anything. He seemed to be alone. I thought it had a landing light on, maybe it was a fire, I don't know.

20:37:51 BOS: Stinger bee 507, roger that and anything else comes to your mind, you can use your other radio; come back to this frequency and tell me about it.

20:37:58 E507: That's all I can think of at this time.

20:38:06 BOS: United 2, Boston on one two four point five two.

20:38:08 UAL002: One two four five two, and is that aeroplane right in front of us now?

20:38:12 BOS: He should be right underneath you. They reported the splashdown right underneath you about, ah, twelve and four miles.

20:38:18 UAL002: It's still burning down there.

20:38:20 BOS: In the water?

20:38:21 UAL002: Well, there's bright red and there's smoke coming up . . . there's fire with smoke.

20:38:30 BOS: Fire with smoke coming out of the water?

20:38:32 UAL002: Right at our position right now;
I can give you a lat/long if you want.
20:38:35 BOS: Absolutely, thank you.
20:38:44 UAL002: It's, ah, forty-thirty nine point one
west zero seven two three eight point zero.
20:38:51 BOS: All right, we got forty thirty-nine point
one west zero seven two three eight point zero.

The CVR of TWA Flight 800 follows:

19:59:50 CAPTAIN: Is the bags back on?
19:59:53 [UNIDENTIFIED VOICE IN COCKPIT]: Yes.
19:59:53 CAPTAIN: Yeah, he was on the whole time.
19:59:56 FLIGHT ENGINEER: Are we reconciled?
19:59:56 [UNIDENTIFIED VOICE IN COCKPIT]: Let's
go.
19:59:58 [UNIDENTIFIED VOICE IN COCKPIT]: Push.
19:59:59 [Sound similar to cockpit door closing]
20:00:01 CAPTAIN: We won't bother telling them that.
You don't mind, huh?
20:00:11 SECOND OFFICER: We'd have a mutiny
back there. Now the lavatories are full.
20:00:15 CAPTAIN: Okay, well she said she'd call me
as soon as they ah –
20:00:16 SECOND OFFICER: Probably have to get the
ATIS [recorded local conditions] now, huh?
20:00:18 FIRST OFFICER: Don't, don't, ah, let them
do their job, Ralph. They'll tell you when they're seated.
20:00:22 ATIS: – visibility greater than one zero ceiling
better than five thousand temperature two eight dew

point two one altimeter three zero zero seven approach in use VOR DME runway 22 left departure runway, runway 22 right and south-west departures runway 31 left from intersection of kilo kilo all pilot are require to read back all runway hold short instructions in interest of noise abatement. Please use the assigned runway. Advise you have tango Kennedy Airport information tango two three five one zulu weather wind two two zero eight visibility –

20:00:36 SECOND OFFICER: All door lights are out.

20:00:37 CAPTAIN: Thank you.

20:00:53 FIRST OFFICER: Tango.

20:01:02 AIRCRAFT GROUND PERSONNEL: Cockpit, ground.

20:01:05 CAPTAIN: Hello, ground.

20:01:06 GROUND: All right, everything is shut down here. You should have all door lights out and when you have clearance you can release the brakes.

20:01:13 CAPTAIN: Yeah, we'll get the clearance. We're waitin' on all the people to sit down. I'll be back with ya in just a second.

20:01:18 [Sound of cabin chime]

20:01:20 GROUND: Okay, we're standing by.

20:01:23 SECOND OFFICER TO FLIGHT ATTENDANT: Hello, darling. Everybody seated? Thanks.

20:01:25 CAPTAIN: Amazing.

20:01:26 SECOND OFFICER: Everybody's seated.

20:01:27 CAPTAIN: Do we have push back clearance to move?

20:01:28 SECOND OFFICER: We're, we're, we cleared to push from FIC or –. No not yet.

77

20:01:31 FIRST OFFICER: You have to call them.
20:01:32 SECOND OFFICER: FIC [FLIGHT INFORMATION CONTROL], TWA 800, gate twenty-seven.
20:01:37 FIC: TWA 800?
20:01:39 FIRST OFFICER: Yeah, we're ready to push.
20:01:42 FIC: TWA 800, you're cleared to push gate twenty-seven.
20:01:46 FIRST OFFICER: Cleared to push.
20:01:47 CAPTAIN: Cleared to push.
20:01:48 SECOND OFFICER: Cleared to push, 800.
20:01:50 CAPTAIN: Okay, ground, we are cleared to push. Yeah, well, wait a minute. Hang on a minute. Did they say everybody was seated? Yeah, they did?
20:01:54 SECOND OFFICER: Yes.
20:01:58 CAPTAIN: Okay, we're cleared to push. Sorry.
20:01:57 GROUND: Brakes released, please.
20:01:58 [Sound similar to parking brake being released]
20:01:59 CAPTAIN: Beacon on, brakes released.
20:02:00 GROUND: Thank you.
20:02:02 FIRST OFFICER: You got something else to do, Ralph?
20:02:05 CAPTAIN: Number one ADP –
20:02:06 FIRST OFFICER: There you go.
20:02:07 CAPTAIN: And the electric?
20:02:08 FIRST OFFICER: It's a command.
20:02:09 CAPTAIN: Electric's on.
20:02:10 FIRST OFFICER: Right. That's a command.
20:02:11 CAPTAIN: Command. Number one ADP on and the electric.

20:02:16 FIRST OFFICER: Before you release the brakes. Block's at oh two, I assume.

20:02:29 SECOND OFFICER: I'm showin' oh two out. Is that what you want?

20:02:32 FIRST OFFICER: That's fine.

20:02:33 CAPTAIN: Yeah.

20:02:34 FIRST OFFICER: That's fine.

20:02:35 CAPTAIN: Okay.

20:02:38 FIRST OFFICER: That's a minute overkill.

20:02:40 CAPTAIN: Yeah, well, that was because they weren't seated. They probably had people standing up and they were . . .

20:02:46 FIRST OFFICER: [Telling them to sit] down.

20:02:50 CAPTAIN: You can bet on it. I still think I'm sittin' too high in this thing.

20:03:00 FIC: Eight hundred.

20:03:11 CAPTAIN: Somebody calling us.

20:03:12 SECOND OFFICER: Go ahead.

20:03:13 FIC: Tell your mechanic to pull you back, push you back far enough so we can get an arrival into your gate.

20:03:18 SECOND OFFICER: Okay.

20:03:21 CAPTAIN: And, ah, ground, FIC wants you to push us back far enough so they can bring somebody in our gate.

20:03:27 GROUND: Okay, we'll do that.

20:03:30 CAPTAIN: Thank you. Ah, there's that, ah, new aeroplane.

20:04:10 FIRST OFFICER: One twenty-nine, yes, sir.

20:04:43 GROUND: Okay, this looks far enough.

20:04:45 CAPTAIN: Okay, if you say so.

20:04:47 GROUND: Brakes parked, please.

20:04:48 [Sound of parking brake being set]

20:04:50 CAPTAIN: Brakes parked.

20:04:51 GROUND: Thank you. Cleared to turn your engines.

20:04:54 CAPTAIN: Okay, we'll turn one, two and four today.

20:04:56 [Sound of two mike clicks]

20:04:59 CAPTAIN: Turn one, please.

20:05:12 FIRST OFFICER: [Sound of cough]

20:05:19 CAPTAIN: Contact.

20:05:22 FIRST OFFICER: You got N–1?

20:05:26 CAPTAIN: I do now.

20:05:27 FIRST OFFICER: You do now.

20:05:29 SECOND OFFICER: Four fifty.

20:05:30 CAPTAIN: It bobbled but not much. And turn [engine] two, please.

20:05:44 FLIGHT ENGINEER: [Contact?]

20:06:00 CAPTAIN: N– one.

20:06:01 SECOND OFFICER: Turning.

20:06:04 CAPTAIN: Two.

20:06:07 SECOND OFFICER: Four hundred.

20:06:24 CAPTAIN: Turn four. [TO GROUND] Turning four.

20:06:27 GROUND: . . . four.

20:06:45 CAPTAIN: Contact.

20:06:51 SECOND OFFICER: Four hundred.

20:06:56 CAPTAIN TO GROUND: Disconnect ground equipment, stand by for hand signals, thank you.

20:06:58 GROUND: Okay.

20:07:13 CAPTAIN: Okay, and after-start checklist when you have a moment.

20:07:20 SECOND OFFICER: Stand by.

20:07:29 CAPTAIN: After-start [checklist].

20:07:30 SECOND OFFICER: After-start checklist. Flight recorder?

20:07:33 CAPTAIN: On.

20:07:34 SECOND OFFICER: Start switches?

20:07:35 CAPTAIN: Off.

20:07:36 SECOND OFFICER: Beacon lights?

20:07:37 CAPTAIN: Are on.

20:07:38 SECOND OFFICER: Brake pressure?

20:07:41 CAPTAIN: Checked.

20:07:42 SECOND OFFICER: Start levers?

20:07:44 CAPTAIN: Idle detent.

20.07:45 SECOND OFFICER: Engine anti-ice?

20:07:46 CAPTAIN: Off.

20:07:50 CAPTAIN: You need to get taxi clearance.

20:07:52 FIRST OFFICER TO GATE HOLD: Kennedy gate hold, TWA's 800 heavy. We're lifeguard, ah, we're ready to taxi out Delta Alpha with Tango.

20:08:01 GATE HOLD: TWA 800, all right contact ground one two one point niner for the taxi inform them that you are lifeguard.

20:08:04 SECOND OFFICER: After-start checklist complete.

20:08:07 FIRST OFFICER: Roger.

20:08:13 FIRST OFFICER TO GROUND CONTROL: Kennedy ground TWA's 800 heavy, lifeguard comin' out Delta Alpha with Tango.

20:08:19 GROUND: Ah, TWA 800 heavy, ah, you're a lifeguard today?

20:08:24 FIRST OFFICER: Yes, sir.

20:08:25 GROUND: You know every day you come out and we don't know that you're a lifeguard and then you tell us you are and, ah, if you could tell company to, ah, you know, ah, put that in their flight plan, ah, it would help us out a lot.

20:08:38 FIRST OFFICER TO GROUND: TWA's 800, understand. I don't think they knew it either until the last minute.

20:08:41 GROUND: All right, TWA 800, taxi right on Alpha and hold short of Echo.

20:08:47 FIRST OFFICER: TWA 800, right Alpha hold short of Echo.

20:08:52 CAPTAIN: Right on Alpha and hold short of Echo. Clear right?

20:08:54 FIRST OFFICER: Clear right.

20.08:55 [Sound of parking break being released]

20:08:57 CAPTAIN: Clear left.

20:09:05 SECOND OFFICER: And LOAD [Passenger/freight Load Controller], TWA 800.

20:09:07 LOAD: 800, stand by.

20:09:19 FIRST OFFICER: Watch number one. It's too high.

20:09:26 CAPTAIN: Forty-five per cent. You got a guy over there.

20:09:30 FIRST OFFICER: Yup.

20:09:34 CAPTAIN: Right on Alpha, huh?

20:09:36 LOAD: 800, ready to copy?

20:09:37 SECOND OFFICER: Ready to copy.

20:09:38 LOAD: On board twenty-nine up front, one-eight-three in the rear, takeoff fuel is one seven six decimal six, your gross takeoff weight is five nine zero seven seven one trim six decimal one and no reported GSIs. Copy?

20:09:41 CAPTAIN: Clear.

20:09:42 FIRST OFFICER: Yeah. Keep it comin'.

20:09:51 CAPTAIN: Does he look clear?

20:09:52 FIRST OFFICER: Yup, it's no problem.

20:09:56 SECOND OFFICER: Okay twenty-nine in the front, one eighty-three in the back, one seven six decimal six on the fuel, five nine zero decimal seven seven one on the takeoff weight, six point one on the trim and no GSIs. TWA 800 out.

20:10:01 CAPTAIN: One two three –. Ya think he's gunna try and get us out being a lifeguard?

20:10:10 LOAD: Okay read back . . . both times?

20:10:12 SECOND OFFICER: Yeah we're out at ah zero zero zero two and, ah, expecting off here probably about, ah, thirty five.

20:10:15 FIRST OFFICER: I think he just . . . That's your undershoot problem, huh.

20:10:21 CAPTAIN: . . . It is.

20:10:24 LOAD: Copy zero two and three five. Have a good flight, 800.

20:10:25 SECOND OFFICER: See ya.

20:10:26 CAPTAIN: How's that look?

20:10:27 FIRST OFFICER: Better.

20:10:50 SECOND OFFICER: Six point one on the trim.

20:10:53 FIRST OFFICER: Okay, set up here.

20:11:58 CAPTAIN: Well we lost a little bit of weight huh? Payload.

20:12:04 GROUND: TWA 800, make a left turn on,

um, taxi way echo behind Carnival and hold short of runway 31 Right and you can monitor tower now on one two three point niner.

20:12:17 FIRST OFFICER: TWA's 800 heavy left echo hold short of three one right over to the tower, bye.

20:12:24 CAPTAIN: Left on echo behind Carnival, hold short of 31 Right.

20:12:41 FIRST OFFICER: Can I have the weight slip if you are done with it, Ollie? Notice that's going to be an undershoot too.

20:13:24 CAPTAIN: What's that?

20:13:25 FIRST OFFICER: Good.

20:13:27 CAPTAIN: Well then someone's given me the wrong poop 'cause I was tryin' to turn, like, on the L ten eleven. They said I was over-turning.

20:13:33 FIRST OFFICER: Ah.

20:13:38 CAPTAIN: How much past centre then?

20:13:40 FIRST OFFICER: Nose wheel is back by the emergency exit door. Right?

20:14:12 TOWER: TWA 800 heavy lifeguard, Kennedy tower.

20:14:13 FIRST OFFICER: TWA's 800 heavy lifeguard, go ahead.

20:14:16 TOWER: I'm gunna put you behind British Airways so the company heavy seven six knows to follow you so make a right on the runway and left at Zulu alpha and follow British.

20:14:23 FIRST OFFICER: TWA 800 heavy, okay right on ah thirty one ah correction thirteen left and follow British.

20:14:29 CAPTAIN: Start the number three motor.

20:14:31 FIRST OFFICER: Let's . . . start taxi.

20:14:36 CAPTAIN: Okay.

20:14:37 FIRST OFFICER: Are you ready?

20:14:38 SECOND OFFICER: Okay.

20:14:39 FIRST OFFICER: Just let me have one engine.

20:14:41 SECOND OFFICER: There you go. All right, we got enough pressure.

20:14:42 FIRST OFFICER: Okay here we go. I'll get the engine for ya.

20:15:43 CAPTAIN: Yeah delayed engine start.

20:15:44 SECOND OFFICER: Delayed engine start checklist. Start switches?

20:15:44 CAPTAIN: Off.

20:15:44 SECOND OFFICER: Start levers?

20:15:45 CAPTAIN: Idle detent.

20:15:47 SECOND OFFICER: Engine anti-ice?

20:15:47 CAPTAIN: Off.

20:15:51 SECOND OFFICER: Delayed engine start checklist is complete.

20:15:53 CAPTAIN: Taxi checklist.

20:15:55 SECOND OFFICER: Taxi checklist. Flaps and runway?

20:15:58 CAPTAIN: Flaps are ten and green for runway 22 Right Kennedy.

20:16:04 SECOND OFFICER: Ten eight green 22 Right Kennedy. Take off data EPR and airspeed bugs?

20:16:08 CAPTAIN: Five hundred and ninety thousand seven seventy one takeoff EPR's set at point three three bugs set and cross checked at one fifty three.

20:16:16 FIRST OFFICER: Set and cross checked.

20:16:17 SECOND OFFICER: Stabilizer trim?

20:16:18 CAPTAIN: Is set at six point one.

20:16:21 SECOND OFFICER: Probe heat?

20:16:22 CAPTAIN: On.

20:16:23 SECOND OFFICER: Flight controls.

20:16:25 CAPTAIN: Checked.

20:16:26 SECOND OFFICER: Auto-brakes?

20:16:28 CAPTAIN: Armed.

20:16:29 FIRST OFFICER: Now you can start it.

20:16:31 SECOND OFFICER: Yaw dampers?

20:16:32 CAPTAIN: On.

20:16:34 FIRST OFFICER: Wrong answer checked.

20:16:35 CAPTAIN: Checked.

20:16:37 FIRST OFFICER: Right here don't roll out start rollin' out you're beside the line.

20:16:43 SECOND OFFICER: Seat belt shoulder harnesses?

20:16:44 CAPTAIN: Checked. Okay, gentlemen, standard TWA crew coordination. You call out eighty, Vee one, Vee R, please.

20:16:58 FIRST OFFICER: That's the first officer's.

20:17:00 CAPTAIN: We're going to fly headings, huh.

20:17:02 FIRST OFFICER: I say that's standard first-officer duties.

20:17:06 CAPTAIN: Well.

20:17:07 SECOND OFFICER: Taxi checklist is complete.

20:17:08 CAPTAIN: Two hundred five degree on the heading, five thousand.

20:17:10 FIRST OFFICER: That's it.

20:17:18 TOWER: TWA 800 heavy, caution wake

turbulence from a 757, Runway 22 Right, taxi into position and hold.

20:17:24 FIRST OFFICER: TWA's 800 heavy lifeguard, position and hold 22 Right.

20:17:28 CAPTAIN: Position and hold 22 Right. [TO SECOND OFFICER]: Will you alert the cabin please.

20:17:40 SECOND OFFICER: Flight attendants please be seated for takeoff.

20:18:03 FIRST OFFICER: Now that's better.

20:18:04 SECOND OFFICER: Now it's coming on . . .

20:18:06 CAPTAIN: I'll just extend it out to that line.

20:18:07 SECOND OFFICER: Sure . . .

20:18:09 FIRST OFFICER: Yeah that's one of the ways you test yourself too is whether when you get rolled out is the whole airplane longitudinally lined up.

20:18:20 TOWER: TWA 800 heavy lifeguard, wind's two four zero at eight runway 22 Right, cleared for takeoff.

20:18:27 FIRST OFFICER: TWA's 800 heavy lifeguard, cleared for takeoff 22 Right.

20:18:31 CAPTAIN: Before takeoff checklist.

20:18:33 SECOND OFFICER: Before takeoff checklist. Icing considerations?

20:18:34 CAPTAIN: Checked.

20:18:35 SECOND OFFICER: Cabin alert?

20:18:36 CAPTAIN: Checked.

20:18:36 SECOND OFFICER: Transponder?

20:18:37 CAPTAIN: That's checked.

20:18:39 SECOND OFFICER: Ignition?

20:18:41 CAPTAIN: Flight start.

20:18:42 SECOND OFFICER: Body gear steering?

20:18:43 CAPTAIN: Disarmed.

20:18:44 FIRST OFFICER: Clocks.

20:18:46 SECOND OFFICER: Before takeoff checklist is complete.

20:18:48 CAPTAIN: Thank you.

20:18:49 FLIGHT ENGINEER: Get right up in there.

20:18:51 [Sound of increasing engine noise]

20:18:59 CAPTAIN: Trim throttles.

20:19:14 FIRST OFFICER: Eighty knots. Vee one. Vee R.

20:19:43 CAPTAIN: Gear up.

20:19:44 FIRST OFFICER: Gear up.

20:19:52 CAPTAIN: That's alive.

20:20:00 TOWER: TWA 800 heavy contact New York departure one three five point niner, good evening.

20:20:05 FIRST OFFICER: TWA's 800 heavy, good night. [TO DEPARTURE] Kennedy departure TWA's 800 heavy leaving nine hundred climbing five thousand.

20:20:19 DEPARTURE: Lifeguard TWA 800 heavy New York departure radar contact climb and maintain one one thousand.

20:20:24 FIRST OFFICER: TWA's 800 heavy climb and maintain one one thousand.

20:20:29 CAPTAIN: Climb to one one thousand and maintain.

20:20:44 DEPARTURE: TWA 800 heavy, turn left heading one five zero.

20:20:47 CAPTAIN: Left to one five zero.

20:20:48 FIRST OFFICER TO DEPARTURE: TWA's 800 heavy, turn left heading one five zero.

20:20:51 CAPTAIN: Flaps five.

20:20:53 FIRST OFFICER: Flaps five.

20:21:11 CAPTAIN: Flaps one.

20:21:12 FIRST OFFICER: Flaps one.

20:21:26 CAPTAIN: Flaps up.

20:21:29 FIRST OFFICER: Say what?

20:21:29 CAPTAIN: Flaps up.

20:21:30 FIRST OFFICER: Flaps up.

20:21:48 CAPTAIN: Climb thrust.

20:22:01 DEPARTURE: TWA lifeguard TWA 800 heavy, turn left heading zero seven zero.

20:22:07 FIRST OFFICER: TWA's lifeguard 800 heavy turn left heading zero seven zero.

20:22:11 CAPTAIN: Left zero seven zero.

20:22:29 DEPARTURE: TWA 800 heavy or lifeguard TWA 800 heavy, turn left heading zero five zero vector climbin' around traffic.

20:22:35 FIRST OFFICER: TWA's 800 heavy, turn left heading zero five zero.

20:22:41 CAPTAIN: Left zero five zero climb vector.

20:22:44 DEPARTURE: TWA 800 heavy, the traffic in the turn will be [at] three o'clock and five miles northeast bound four thousand nor- is a company seven two five five in trail will be a Saab-Fairchild. When you're out of five I'll have on course.

20:22:54 FIRST OFFICER: TWA's 800 heavy understand.

20:22:58 CAPTAIN: He's at three o'clock?

20:23:00 FIRST OFFICER: Yeah. That's the problem.

20:23:19 DEPARTURE: TWA 800 heavy direct Betty resume own navigation.

20:23:22 FIRST OFFICER: TWA's 800 heavy direct Betty own navigation.

20:23:26 CAPTAIN: Direct Betty and our own nav.

20:23:37 DEPARTURE: TWA lifeguard TWA 800 heavy contact Boston one three two point three.

20:23:38 FIRST OFFICER: Huh.

20:23:39 CAPTAIN: Direct Betty. Correct?

20:23:42 FIRST OFFICER: TWA's 800 heavy, ah, say again the frequency.

20:23:44 DEPARTURE: One three two point three.

20:23:46 FIRST OFFICER: TWA's 800 heavy, good day.

20:24:30 CAPTAIN: Seems like a homesick angel here . . . awesome.

20:24:36 FIRST OFFICER: It's bleeding off airspeed, that's why.

20:24:38 CAPTAIN: Yeah . . .

20:24:41.7 FIRST OFFICER: New York centre TWA's lifeguard 800 heavy eight thousand two hundred climbing one one thousand.

20:24:48 CONTROL: TWA 800 Boston centre, roger, climb and maintain one three thousand.

20:24:53.4 FIRST OFFICER: TWA's 800 heavy climb and maintain one three thousand.

20:24:57 CAPTAIN: Climb and maintain one three thousand.

20:25:31 CONTROL: TWA 800 what's your rate of climb?

20:25:34.5 FIRST OFFICER: TWA's 800 heavy, ah, about two thousand feet a minute here until accelerating out of ten thousand.

20:25:41 CONTROL: Roger, sir, climb and maintain flight level one niner zero and expedite through fifteen.

20:25:47.1 FIRST OFFICER: TWA's 800 heavy climb and maintain one niner zero and expedite through one five thousand.

20:25:53 CAPTAIN: Climb to one nine zero expedite through one five thousand.

20:25:57 SECOND OFFICER: Pressurization checks. [Takeoff] thrust go on cross feed?

20:26:04 CAPTAIN: Yeah.

20:26:07 SECOND OFFICER: I'll leave that on for just a little bit. Is that right?

20:26:13 FLIGHT ENGINEER: Yes.

20:26:24 CONTROL: TWA 800, amend the altitude maintain, ah, one three thousand thirteen thousand only for now.

20:26:29 CAPTAIN: Thirteen thousand.

20:26:30.3 FIRST OFFICER TO CONTROL: TWA 800 heavy, okay, stop climb at one three thousand.

20:26:35 CAPTAIN: Stop climb at one three thousand.

20:26:59 FIRST OFFICER: Twelve for thirteen.

20:27:35 [Sound of click].

20:27:47 [Sound of altitude alert tone].

20:28:13 CONTROL: TWA 800 you have traffic at one o'clock and, ah, seven miles south bound a thousand foot above you. He's, ah, Beech nineteen hundred.

20:28:20.6 FIRST OFFICER: TWA 800 heavy, ah, no contact.

20:28:25.7 SECOND OFFICER: 800 with an off report, ah, plane number one seven one one nine. We're out at zero zero zero two, and we're off at zero zero one nine, fuel one seven nine decimal zero, estimating Charles De Gaulle at zero six two eight.

20:28:42 TWA FLIGHT INFORMATION
CONTROLLER: TWA 800, got it all.
20:28:44.8 SECOND OFFICER: Thank you.
20:29:15 CAPTAIN: Look at that crazy fuel flow
indicator there on number four. See that? Somewhere in
here I better trim this thing (in/up).
20:29:39 FIRST OFFICER: Huh?
20:29:39 CAPTAIN: Someplace in here I better find out
where this thing's trimmed.
20:30:15 CONTROL: TWA climb and maintain one
five thousand.
20:30:18 CAPTAIN: Climb thrust.
20:30:19.2 FIRST OFFICER: TWA 800 heavy climb
and maintain one five thousand, leaving one three
thousand.
20:30:24 CAPTAIN: Ollie.
20:30:24 SECOND OFFICER: Huh?
20:30:25 CAPTAIN: Climb thrust. Climb to one five
thousand.
20:30:35 SECOND OFFICER: Power's set.
20:30:42 [Sound similar to a mechanical movement in
cockpit]
20:31:05 [Sounds similar to recording tape damage
noise]
[The CVR recorded a very loud sound for a fraction of a
second (0.117 second) on all channels immediately before
the recording ended.]
20:31:12 END OF RECORDING

All 212 passengers and the crew of four and the cabin crew
of 14 died in the explosion.

The NTSB determined that the probable cause of the TWA flight 800 accident was an explosion of the centre wing fuel tank (CWT), resulting from ignition of the flammable fuel/air mixture in the tank. The source of ignition energy for the explosion could not be determined with certainty, but, of the sources evaluated by the investigation, the most likely was a short circuit outside of the CWT that allowed excessive voltage to enter it through electrical wiring associated with the fuel quantity indication system.

Contributing factors to the accident were the design and certification concept that fuel tank explosions could be prevented solely by precluding all ignition sources and the design and certification of the Boeing 747 with heat sources located beneath the CWT with no means to reduce the heat transferred into the CWT or to render the fuel vapour in the tank non-flammable.

MONROE, Michigan, USA
9 January 1997

At 3.08 p.m. on 9 January 1997, an Empresa Brasileira de Aeronautica, S/A EMB-120RT, a twin-propeller-driven aircraft registered as N265CA, operated by COMAIR Airlines, Inc., as Flight 3272, left the Cincinnati/Northern Kentucky International Airport, carrying twenty-six passengers, one flight attendant and a cockpit crew of two. The flight was travelling to Detroit, Michigan, under instrument meteorological conditions and instrument flight rules.

15:23:49 INDIANAPOLIS CONTROL: Comair 3272 climb maintain flight level two one zero [21,000 feet].
15:23:52 CAPTAIN TO CONTROL: Flight level two one zero, Comair 3272, thank you.
15:23:57 FIRST OFFICER: Twenty-one.
15:23:59 CAPTAIN: Set once and set twice.
15:25:44 [Sound of three tones similar to that of the altitude alerter]
15:26:59 INDIANAPOLIS CONTROL: Comair 3272, any improvements, ah, with the climb there?
15:27:03 CAPTAIN: Comair 3272, affirmative . . . it's, ah, smooth here at two one oh . . . we're getting, ah, occasional light chop at one nine oh 'cause we were right at the tops.

15:27:11 INDIANAPOLIS CONTROL: Right at the tops . . . appreciate it, thanks.

15:29:09 [Sound similar to that of slight decrease in propeller rpm frequency] [Sound of several unknown clicks]

15:29:52 CAPTAIN: Power and NP set, pressurization check, cruise check complete.

15:31:22 INDIANAPOLIS CONTROL: Comair 3272, contact Cleveland one two three point niner.

15:31:28 CAPTAIN: One two three point niner, Comair 3272, good day.

15:31:31 INDIANAPOLIS CONTROL: So long.

15:31:32 [Sound of tone similar to that of frequency change]

15:31:34 CAPTAIN: Good afternoon Cleveland Centre, Comair 3272 . . . flight level two one zero.

15:31:39 CLEVELAND: Comair 3272, Cleveland Centre, roger. Comair 3272, how's your ride there?

15:33:08 CAPTAIN: Comair 3272, it's smooth.

15:35:37 CLEVELAND: Comair 3272, descend and maintain one two thousand . . . the, ah, Detroit altimeter two nine two one.

15:35:44 CAPTAIN: Two nine two one . . . descend and maintain one two twelve thousand, Comair 3272.

15:35:52 FIRST OFFICER: Yeah, there's twelve . . . gotta go [sound of human whistling].

15:35:57 CAPTAIN: Down.

15:36:07 CLEVELAND: Comair 37, I'm sorry . . . 3272, contact Cleveland one two zero point four five.

15:36:13 CAPTAIN: One two zero point four five Comair 3272, good day.

15:36:43 FIRST OFFICER: [We] might get a speed warning here.

15:36:46 CAPTAIN TO CLEVELAND CONTROL: Good afternoon, Cleveland Centre, Comair 3272 ... flight level one nine oh ... descend and maintain one two twelve thousand.

15:36:52 CLEVELAND: 3272, Cleveland Centre, roger ... no delay down to twelve for traffic.

15:36:56 CAPTAIN: Comair 3272, wilco.

15:37:04 FIRST OFFICER: Dive, dive.

15:37:09 FIRST OFFICER: Whoop, whoop dive [sound of human whistling].

15:38:03 [Sound of chime similar to that of flight attendant chime]

15:38:07 FLIGHT ATTENDANT TO CAPTAIN: Need anything?

15:38:14 CAPTAIN: Would you be kind enough to get me an ice refresher ... just add some to that.

15:38:18 FIRST OFFICER: No thanks ... doing great.

15:38:20 FLIGHT ATTENDANT: Are we already on the descent?

15:38:21 CAPTAIN: Yes, ma'am ... You told them an hour but it's only [going to be] forty minutes today.

15:38:27 FLIGHT ATTENDANT: Forty minutes.

15:38:27 CAPTAIN: Didn't I ... I think I told you forty-five.

15:38:30 FLIGHT ATTENDANT: Oh that's right. You did it.

15:38:32 CAPTAIN: I go okay ... well, she just knows we're gonna probably have to hold now.

15:38:35 [Sound of laughter]

15:38:55 CAPTAIN: Yah, we're only seventy-five miles out right now.

15:38:57 FLIGHT ATTENDANT: Okay, great, thanks.

15:38:58 CAPTAIN: So –

15:38:58 FLIGHT ATTENDANT: Here's your drink.

15:39:00 CAPTAIN: Fifteen . . . fifteen minutes about.

15:39:03 FLIGHT ATTENDANT: Okay.

15:39:04 CAPTAIN: Thank you very much for the ice.

15:39:07 CLEVELAND: Comair 3272, continue descent to one one thousand then fly heading of, ah, zero three zero to rejoin the MIZAR arrival off of Detroit.

15:39:14 CAPTAIN TO CONTROL: Descend and maintain one one eleven thousand and heading zero three zero to join the arrival, Comair 3272.

15:39:24 FIRST OFFICER: Okay, there's, ah, there's eleven and out of Detroit we're looking at, ah . . . two thirty-nine radial.

15:40:09 RECORDED CONDITIONS DETROIT, PICKED UP ON CVR: Detroit metropolitan airport information hotel . . . two zero two six Zulu special . . . wind zero seven zero at six . . . visibility one . . . light snow . . . six hundred scattered . . . ceiling one thousand four hundred broken . . . two thousand one hundred overcast . . . temperature minus three . . . dew point minus four . . . altimeter two niner two one . . . remarks . . . A02 . . . tower visibility one and one half . . . papa zero zero zero zero . . . ILS approach in use runway three right . . . departing runway three centre . . . notices to airmen . . . runway two one right, three left closed . . .

runway two seven left, nine right closed . . . runway two seven right, nine left closed . . . taxiway Yankee eleven closed . . . braking action advisories in effect . . . (local) de-ice procedure in effect . . . gate hold procedure in effect for Newark, Kennedy, Chicago-O'Hare, Philadelphia, St Louis, GRR airports . . . advise on initial contact you have information hotel.

15:40:35 CAPTAIN: So what do you got set up here?

15:40:38 FIRST OFFICER: I've got Detroit. Let's see which one of those lines we hit first.

15:40:48 [Sound of three tones similar to that of the altitude alerter]

15:40:49 CAPTAIN: Okay.

15:41:16 [Sound of audio interrupt similar to that of tape splice]

15:42:43 [Sound of yawn]

15:42:51 CLEVELAND: Comair 3272, Detroit one two four point niner seven.

15:42:54 CAPTAIN: One two four point niner seven Comair 3272, good day. [Frequency change] Good afternoon Detroit approach . . . Comair 3272, one one eleven thousand hotel.

15:43:07 DETROIT APPROACH: Comair 3272, Detroit approach . . . depart MIZAR heading zero five zero vector to ILS runway 3 right final approach course . . . runway 3 right braking action reported poor by a DC niner.

15:43:16 CAPTAIN: Roger depart MIZAR heading zero five zero, Comair 3272.

15:44:11 DETROIT APPROACH: Comair 3272, maintain one niner zero knots . . . if unable advise.

15:44:15 CAPTAIN: Roger, one niner zero knots, Comair 3272.

15:44:50 FIRST OFFICER: And five miles for ah . . . for ah MIZAR.

15:45:46 DETROIT APPROACH: Comair 3272, descend and maintain seven thousand.

15:45:49 CAPTAIN: Seven thousand Comair 3272. [TO FIRST OFFICER] And seven's in the altitude alerter.

15:45:56 FIRST OFFICER: Seven's verified . . . there's MIZAR and we're turning zero five zero.

15:46:12 DETROIT: Comair 3272 turn left heading zero three zero vector for sequencing.

15:46:14 CAPTAIN: Zero three zero Comair 3272.

15:46:57 DETROIT: North-west 272 information Alpha is current . . . winds zero six zero at six . . . visibility one and one half . . . light snow . . . ceiling six hundred broken one thousand one hundred broken . . . two thousand one hundred overcast . . . altimeter's two niner one niner . . . runway three right RVR three thousand five hundred . . . braking action reported poor by DC niner. Comair 3272, turn right heading zero five five.

15:47:21 CAPTAIN: Zero five five Comair 3272.

15:47:32 FIRST OFFICER: [Sound similar to that of a human sniffle]

15:47:47 FIRST OFFICER: Let's run the descent check.

15:47:49 CAPTAIN: Ice protection?

15:47:51 FIRST OFFICER: Windshield, props, standard seven.

15:47:53 CAPTAIN: Ignition?

15:47:54 FIRST OFFICER: Auto.

15:47:55 CAPTAIN: Pressurization?

15:47:55 FIRST OFFICER: It's, ah, reset for landing in Detroit . . . six thirty-nine . . . looks good.

15:48:01 CAPTAIN: Altimeters?

15:48:03 FIRST OFFICER: Ah, twenty-one.

15.48:04 CAPTAIN: Set left.

15:48:05 FIRST OFFICER: Set right.

15:48:06 CAPTAIN: Landing lights?

15:48:06 FIRST OFFICER: Landing lights on.

15:48:07 CAPTAIN: Seat belt sign?

15:48:08 FIRST OFFICER: On.

15:48:09 CAPTAIN: Packs and bleeds?

15:48:10 FIRST OFFICER: Four lows.

15:48:11 CAPTAIN: Cross feed?

15.48:12 FIRST OFFICER: Cross feed's off.

15.48:12 CAPTAIN: That completes that.

15:48:14 FIRST OFFICER: Okay . . . a thousand to go . . . Uhm . . . we're going to do an ILS to runway 3 right . . . It'll be a coupled approach . . . flaps twenty-five . . . frequency is one one one point five . . . that's set . . . might as well set that in there . . . inbound course is zero three five.

15:48:38 [Sound of three tones similar to that of the altitude alerter]

15:48:43 FIRST OFFICER: We're gonna intercept the top somewhere, ah, whatever altitude he gives us . . . ah . . . twenty-seven hundred's the intercept to the glide slope.

15:48:47 DETROIT APPROACH: Comair 3272, turn right heading zero seven zero.

15:48:50 CAPTAIN: Zero seven zero Comair 3272.

15:48:55 [Sound similar to increase in engine/prop noise]

15:49:13 FIRST OFFICER: It's, ah, two-hundred-foot approach with the decision altitude of eight thirty-three . . . you've already got that set in there . . . missed approach will be published climb to eleven hundred . . . and a climbing right turn to three thousand direct to the, ah, DM locator outer marker 'Spencer' . . . which is, ah, two twenty-three that's set . . . and hold . . . that will be a teardrop entry . . . questions, comments.

15:49:38 CAPTAIN: No questions . . . twenty-one, fourteen, and forty-three are your bugs.

15:49:43 FIRST OFFICER: Twenty-one, fourteen, forty-three . . . set.

15:49:48 CAPTAIN: Autofeather?

15:49:50 FIRST OFFICER: Armed.

15:49:51 CAPTAIN: Nav radios.

15:49:53 FIRST OFFICER: Nav radios are, ah, set eleven point five.

15:49:54 DETROIT: Comair 3272, turn right to a heading of one four zero . . . reduce speed to one seven zero.

15:49:59 CAPTAIN: Heading one four zero speed one seven zero Comair 3272.

15:50:11 CAPTAIN: I'll be on two.

15:50:11 FIRST OFFICER: All right.

15:50:15 CAPTAIN: Good afternoon, Detroit . . . Comair 3272.

15:50:28 CAPTAIN TO FIRST OFFICER: Nobody likes to answer me . . . I'm back.

15:50:28 DETROIT: Comair 3272, contact approach one two five point one five so long.

15:50:32 CAPTAIN: One two five point one five Comair 3272, good day.

15:50:36 [Sound of tone similar to that of frequency change]

15:50:38 FIRST OFFICER: Maybe you should try being abusive with them.

15:50:40 CAPTAIN: Huh?

15:50:41 FIRST OFFICER: Gotta try being abusive with them.

15:50:43 CAPTAIN: That's right.

15:50:43 FIRST OFFICER: Answer the phone, dummy.

15:50:44 CAPTAIN: Yeah.

15:50:45 CAPTAIN: Good afternoon, Detroit approach, Comair 3272, seven thousand.

15:50:49 DETROIT: Comair 3272, Detroit approach . . . reduce speed to one seven zero and maintain six thousand.

15:50:54 CAPTAIN: Speed one seven zero . . . descend and maintain six thousand, Comair 3272.

15:50:57 [Sound of three tones similar to that of the altitude alerter]

15:51:00 CAPTAIN: Six.

15:51:00 FIRST OFFICER: Six thousand.

15:51:14 DETROIT: Comair 3272, fly heading one four zero.

15:51:17 CAPTAIN: One four zero, Comair 3272.

15:51:20 FIRST OFFICER: Wonder what plane he's looking at?

15:51:25 CAPTAIN: Ah, the one, ah, that's not going one four zero.

15:51:27 AIRLINE OPERATIONS: 3272, you calling Detroit?

15:51:30 CAPTAIN: Yes, sir, we're in range . . . ah, positive fuel . . . We'll be at the gate in approximately nine minutes and forty-eight seconds.

15:51:38 OPERATIONS: Approximately, huh?

15:51:39 CAPTAIN: Approx, of course.

15:51:41 OPERATIONS: Do you guys know if you have any special assistance coming in?

15:51:44 CAPTAIN: I can't recall anyone coming out so I thi– I think we're all good there . . . all we'll need is just fuel.

15:51:51 OPERATIONS: Roger that, ah, bravo three.

15:51:53 [Sound of three tones similar to that of the altitude alerter]

15:51:54 CAPTAIN TO OPERATIONS: Roger that . . . and how many do we have going back so I know how many seats to give you?

15:51:58 OPERATIONS: Ah, you're booked to twenty-eight right now.

15:52:01 CAPTAIN: Okay, that'll be our load then . . . We can take them all . . . thanks.

15:52:07 CAPTAIN TO FIRST OFFICER: Took 'em a while but they came back to me.

15:52:13 FIRST OFFICER: That's good news . . . No changes while you were away.

15:52:13 DETROIT: Comair 3272, descend and maintain four thousand.

15:52:16 CAPTAIN: Four thousand, Comair 3272. Four's in the altitude alert.

15:52:23 FIRST OFFICER: Four thousand verified.

15:53:03 DETROIT: Cactus fifty [AMERICA WEST FLIGHT 50] zero six zero to intercept three right.

15:53:05 AMERICA WEST FLIGHT 50: Zero six zero okay . . . you got any wind shear reports on the final?

15:53:09 DETROIT: Ah, no . . . I haven't had anything, ah . . . just, ah, slick runways and, ah, low visibilities.

15:53:15 AMERICA WEST FLIGHT 50: Okay . . . yeah, it's two thirty-seven at thirty-two up here.

15:53:18 DETROIT: Ah, you'll pick up a headwind once you get down, ah, probably, oh, two thousand feet or so.

15:53:25 DETROIT: Comair 3272, turn right heading one eight zero . . . reduce speed to one five zero.

15:53:29 CAPTAIN: Heading one eight zero . . . speed one five zero, Comair 3272.

15:53:42 [Sound of three tones similar to that of the altitude alerter]

15:53:42 DETROIT: Comair 3272, reduce speed to one five zero.

15:53:45 CAPTAIN: Speed one five zero, Comair 3272.

15:53:52 FIRST OFFICER: This guys got –

15:53:53 CAPTAIN: They gotta always tell us twice.

15:53:55 FIRST OFFICER: He's got short-term memory disorder, I think.

15:53:57 CAPTAIN: Is that what that is?

15:53:58 FIRST OFFICER: Yeah, he's got Alzheimer's . . . that's what it is.

15:53:59 DETROIT: Comair 3272 now turn left heading zero nine zero . . . plan a vector across the localizer.

15:54:04 CAPTAIN: Heading zero niner zero Comair 3272.

15:54:14.4 [Sound of several unidentified thumps fade in and out]

15:54:15.9 [Sound of several 'whirring' noises, similar to that of the elevator trim servo]

15:54:16.0 [Sound of increase in discrete high frequency noise similar to that of power increase]

15:54:17.1 [Significant reduction in background ambient noise]

15:54:20.8 CAPTAIN: Looks like your low-speed indicator.

15:54:23.6 CAPTAIN: Power.

15:54:23.9 [Sound similar to that of stick shaker starts]

15:54:24.1 FIRST OFFICER: Thanks.

15:54:24.1 [Sound of three chimes and autopilot aural warning]

15:54:25.9 [Sound similar to that of stick shaker stops]

15:54:26.1 FIRST OFFICER: Oh.

15:54:26.1 CAPTAIN: Oh, shit.

15:54:26.8 [Sound of increase in background noise similar to that of power increase]

15:54:29.0 [Sound of ground proximity warning system 'bank angle' aural warning]

15:54:29.1 [Sound of three chimes and autopilot aural warning]

15:54:31.0 [Sound similar to that of stick shaker starts and continues to the end of tape]

15:54:33.3 [Sound of single intake of human breath]

15:54:34.3 [Sound of three chimes and autopilot aural warning]

15:54:35.3 [Sound of ground proximity warning system 'bank angle' aural warning]

15:54:37.1 CAPTAIN: [Exclamation]
15.54:38.2 [Sound of three chimes and autopilot aural warning]
15:54:39.1 CAPTAIN: [Exclamation]
15:54:40.1 END OF RECORDING

At 15:54:26.1 the CVR recorded the first officer and the captain stating respectively, 'Oh' and 'Oh, shit'. According to the flight data recorder, the aeroplane's left-roll attitude was increasing to more than 140 degrees, and the pitch attitude was decreasing to nearly 50 degrees nose down by 15.54:29. At 15:54:29.0 the ground proximity warning system (GPWS) *'bank angle'* aural warning sounded, followed by three chimes and the autopilot aural warning; these warnings sounded repeatedly as the aircraft descended to the ground. The aeroplane struck the ground in a steep nose-down attitude in a level field in a rural area about nineteen miles south-west of the Detroit International Airport. The NTSB determined that the probable cause of the accident was icing on the wings which led to the loss of control when the aeroplane formed a thin, rough accretion of ice on its lifting surfaces. All twenty-nine people on the aircraft died.

NANTUCKET, Massachusetts, USA

31 October 1999

Because of the ten-hour scheduled flight time from New York to Cairo, EgyptAir Flight 990 carried two designated flight crews (each crew consisting of a captain and first officer). The flight took off at 1.01 a.m. from New York JFK's runway 22R and climbed to 33,000 feet. At about 1.40 (forty minutes after takeoff), as the aeroplane was climbing to its assigned altitude, the relief first officer offered to relieve the command first officer at the controls. He said, 'I'm not going to sleep at all. I might come and sit for two hours, and then . . .' In other words, he wanted to fly his part of the trip at that time. The command first officer said, 'But I slept. I slept . . .'

The relief first officer said, 'You mean you're not going to get up? You will get up, go and get some rest and come back.'

The command first officer then said, 'You should have told me. You should have told me this, Captain Gamil. You should have said [that the relief first officer] will work first. Just leave me a message. Now I am going to sit beside you. I mean, now, I'll sit by you on the seat. I am not sleepy. Take your time sleeping and when you wake up, whenever you wake up, come back, Captain.' The relief first officer replied, 'I'll come either way. Come work the last few hours, and that's all.' The command first officer

107

responded, 'No, that's not the point. It's not like that. If you want to sit here, there's no problem.' The relief first officer said, 'I'll come back to you, I mean, I will eat and come back, all right?' The command first officer replied, 'Fine. Look here, sir. Why don't you come so that . . . You want them to bring your dinner here, and I'll go to sleep [in the cabin]?' The relief first officer said, 'That's good.' The command first officer then stated to the command captain, 'With your permission, Captain?'

At 01:40:56 the cockpit door opened. Almost a second later, the command first officer stated in a soft voice, 'Do you see how he does whatever he pleases?'

At 01:41:09 the command first officer said, 'No, he does whatever he pleases. Some days he doesn't work at all.'

At 01:41:51 the command first officer left his seat and the relief first officer moved into his seat.

01:36:42 CAPTAIN: What's your opinion about the crowding on this plane?
01:36:44 FIRST OFFICER: What crowding?
01:36:47 CAPTAIN: The crew that is aboard this plane.
01:36:50 FIRST OFFICER: There is one [replacement crew member] going back that has simulator and another whom I have told in Egypt to come back on the same aeroplane from Los Angeles direct to Egypt for training . . . What difference does it make?
01:37:00 FIRST OFFICER: What difference does it make?
01:37:01 CAPTAIN: No, it doesn't make any difference to me. I am saying that for their sake . . .

01:37:07 FIRST OFFICER: . . . because those, you see, are not active.

01:37:08 CAPTAIN: I know that. I didn't say anything. But I mean when you are surprised to see people about whom you don't know anything. It's good that you are here, you know. I don't know, I mean, anybody could . . .

01:37:24 FIRST OFFICER: Isn't the extra crew written in the general dec?

01:37:28 CAPTAIN: I mean, as long as you are here, there is no problem. You are here. If you were not here, anyone could tell me, 'Captain Hatem, should I say okay?'

01:37:38 FIRST OFFICER: No, I sent a telex concerning this operation.

01:37:40 CAPTAIN: Bravo for you, bravo for you, bravo for you, and what about the one that doesn't have?

01:37:45 FIRST OFFICER: What do you mean by the 'one that doesn't have'?

01:37:46 CAPTAIN: The one that comes aboard and doesn't have any document that says that he should board.

01:37:49 FIRST OFFICER: They're all reported to the station from Egypt, each with his own schedule. Anyone who will come back should inform the station in Egypt.

01:37:57 CAPTAIN: Good, what about the one who comes back and nobody knows anything about him?

01.38:01 FIRST OFFICER: I gave my instructions in the station . . .

01:38:05 CAPTAIN: Fine. That's what I told them. I'm not a troublemaker. I mean, you know me. I don't need any headache. I'm telling them I am looking after all your interests. When you all are covered, neither the station nor any pilot can argue with you all, and should anything happen to you, you see, nobody could say I know nothing about you.

01:38:12 [Sound similar to cockpit door operating]

01:38:28 FIRST OFFICER: Why, why would I put anybody, Habashi?

01:38:31 CAPTAIN: No, that's what I'm saying.

01:38:32 FIRST OFFICER: Even if somebody makes a mistake, we'll cover him, too.

01:38:35 CAPTAIN: I'm not saying anything. I'm not saying anything but I get surprised by three, four people going back [to Cairo on this trip] and I know nothing. Everyone is telling me that [the relief first officer] has work in Egypt . . .

01:38:50 FIRST OFFICER: May God make the seven six prosperous, Captain Habashi. Tomorrow, we'll have a hard time up there. Let us be, 'Uncle Haj'.

01:38:56 CAPTAIN: Why, 'Uncle', by God?

01:38:58 FIRST OFFICER: Very sweet, may our Lord make it prosperous. Sweet, sweet.

01:39:00 CAPTAIN: Have we ever shorted you?

  [Sound of laughter]

01:39:01 FIRST OFFICER: That's why I'm saying to you, sir, may our Lord make it prosperous, but we'll have a hard time . . .

01:39:04 CAPTAIN: This man is saying, he's saying to you, 'Even the one that makes a mistake, I will cover him.' I mean, where can you see someone who says that?

01:39:11 FIRST OFFICER: So, why are you upset, Captain?

01:39:12 CAPTAIN: I'm not upset, I'm only speaking because I'm actually finding people about whom I know nothing. That's why they got upset with me.

01:39:21 FIRST OFFICER: The man didn't get upset or anything.

01:39:22 CAPTAIN: No, I'm not talking about those. When I say, 'People, please, anyone not scheduled on the plane should have something saying that he should [be on] board the plane', am I wrong in saying that? But I didn't tell anybody to board or not to board or, 'You have it' or 'You don't have it.'

01:39:40 FIRST OFFICER: Nobody can go back without receiving instructions.

01:39:43 CAPTAIN: That's what I'm telling you . . .

01:39:45 FIRST OFFICER: Nobody can go back like this on his own. Nobody can go back without receiving instructions.

01:39:51 [Sound similar to cockpit door operating]

01:39:56 CAPTAIN: How are you, Jimmy?

01:39:58 RELIEF FIRST OFFICER: How are you, sir?

01:39:59 CAPTAIN: What's new?

01:39:59 RELIEF: I slept, I swear. I, I slept. I slept. You mean you're not going to get up? You will get up, go and get some rest and come back.

01:40:14 FIRST OFFICER: You should have told me. You should have told me this, Captain Gamil. You should have said, 'Abdul . . . I will work first'.

01:40:16 RELIEF: Did I even see you?

01:40:16 FIRST OFFICER: Just leave me a message. Now I am going to sit beside you. I mean, now, I'll sit by you on the seat. I am not sleepy. Take your time sleeping and when you wake up, whenever you wake up, come back, DS [captain].

01:40:28 RELIEF: I'll come either way . . . come work the last few hours, and that's all.

01.40:28 FIRST OFFICER: No . . . that's not the point. It's not like that. If you want to sit here, there's no problem.

01:.40:37 RELIEF: I'll come back to you, I mean, I will eat and come back, all right?

01.40:42 [Sound similar to cockpit door operating]

01:40:43 FIRST OFFICER: Fine, look here, sir. Why don't you come so that . . . you want them to bring your dinner here, and I'll go sleep.

01:40:49 RELIEF: That's good.

01:40:52 FIRST OFFICER: With your permission, Captain.

01:40:56 [Sound similar to cockpit door operating]

01:40:57 FIRST OFFICER: [Spoken in a soft voice to captain] Do you see how he does whatever he pleases?

01:40:59 CAPTAIN: Do you know why that is? That's because you all get upset with me.

01:41:03 FIRST OFFICER: I'm not upset with you, what is it to me?

01:41:03 CAPTAIN: I don't mean you specifically, Abdul. Son. You are . . . look, are you a youngster?

01:41:09 FIRST OFFICER: No, he does whatever he pleases. Some days he doesn't work at all.

01:41:12 CAPTAIN: That's why I'm saying. You see, it's just like you heard. When I told him about those who want to go back, what did he say? He said, 'Never mind, I have them covered.'

01:41:20 [Sound similar to cockpit door operating]

01.41:24 CAPTAIN: The conversation just happened in front of you.

01:41:28 FIRST OFFICER: Doesn't he want to work with Raouf, or what?

01:41:30 CAPTAIN: It's possible, it's possible, God knows. Look, you don't have a male or female camel tied up in this situation [figuratively meaning: you have no personal interest at stake], as they say. Right? By the Prophet, he's just talking nonsense.

01:41:44 [Sound similar to cockpit door operating]

01:41:52 CONTROL: EgyptAir 990.

01:41:57 FIRST OFFICER: Go ahead, 990.

01:42:00 CONTROL: EgyptAir 990, you're cleared to Hotel, Echo, Charlie, Alpha, via after DOVEY, NAT TRACK Zulu, SANTIAGO. Maintain flight level three three zero. maintain Mach point eight zero.

01:42:13 FIRST OFFICER: EgyptAir 990, cleared to Cairo Zulu, TRACK Zulu, SANTIAGO three three zero, eight zero Mach, TRACK identification, three zero four.

01:42:24 CONTROL: EgyptAir 990, readback correct.

01:42:31 FIRST OFFICER: Everything's under control.

01:42:32 RELIEF FIRST OFFICER: What?

01:42:33 FIRST OFFICER: Everything's under control, Haj.

01:42:34 RELIEF: Okay, chief.

Now, about twenty-two minutes after takeoff, the relief first officer has already suggested that he relieve the command first officer. A transfer of control early in the flight was contrary to EgyptAir practice; it was standard for flight crews to wait until three or four hours into the flight before relieving the command crew members. The command first officer had initially reacted with surprise and resistance to the relief first officer's suggestion that he assume control at that time, indicating that the offer was unexpected. However, after some discussion, the command first officer agreed to the change, and at around 01.42 the command first officer vacated his position and the relief first officer moved into his right seat.

01:42:35 CAPTAIN: Thanks, Abdul. [TO RELIEF FIRST OFFICER, NICKNAMED 'JIMMY,' NOW IN THE RIGHT SEAT HAVING REPLACED THE ORIGINAL FIRST OFFICER, ABDUL] How are you, Jimmy?

01:43:00 RELIEF: Why did you bring up this subject again?

01:43:04 CAPTAIN: Look, it's the same story with these number threes . . . [reference to additional repositioning crew members] . . . so when they fly as active crew . . . what are they going to do? You tell me.

01:43:12 RELIEF: It's disorganized.

01:43:13 CAPTAIN: No, that's why I mentioned it to him. Weren't you sitting there listening to me?

01:43:16 RELIEF: I mean, you don't . . . don't . . . he's trying to please them. Just so you know.

01:43:22 CAPTAIN: Oh yes, didn't he say if somebody made a mistake . . .

01:43:24 RELIEF: Yeah.

01:43:25 CAPTAIN: You know, I don't harm anyone, I will cover for him.

01:43:31 CAPTAIN: Did you give the instructions that they should come aboard? He said to me, 'I always send telex messages.' That's it.

01:43:38 RELIEF: He's trying to please them. As long as he pleases them, don't worry your head.

01:43:52 CAPTAIN: Since when do I worry my head . . . ? I mean, you know.

01:43:56 RELIEF: They, they . . . the word is that [you are] making trouble. You are making trouble. That's what's being said by everyone. Just so you know.

01:44:06 CAPTAIN: I don't care.

01:44:09 RELIEF: These guys are a bunch of shits. And what's more, they are being controlled by the guy named, under the leadership of, of the guy named, the shit named [name withheld], you know? Don't worry your head. You're good-hearted.

01:44:26 CAPTAIN: I mean . . . you know me, it doesn't make any difference to me. But when kids start that way . . . I told him, I told him later that if I saw anyone aboard that didn't get permission from anybody and who also keeps going back and forth [to and from Egypt], even with permission, he won't fly with me. I won't fly with him, I'm not willing. I don't want to fly with anyone, even if it comes to me not flying, and if they can prevent me from flying, that's fine. What can I say to them?

01.45:04 RELIEF: These kids are forming a clique with each other, just so you know, under the leadership of [name withheld] . . . [he] controls that group. He has [name withheld]'s ear, as well as . . . This kid is clever and cunning.

01:45:35 RELIEF: Don't you have the flight report with you, Abdul?

01:45:37 FIRST OFFICER: I've got it, sir, I've got it.

01:45:41 CAPTAIN: What's with you? Why did you get all dressed in red like that? When do you have simulator?

01:45:49 RELIEF CREW MEMBER WHO ENTERS COCKPIT: Wednesday.

01:45:50 CAPTAIN: What's today?

01:45:51 CREW MEMBER: Sunday . . . Saturday.

01:45:59 CAPTAIN: Hey, guy, why didn't you take tomorrow's plane?

01:46:00 ANOTHER RELIEF CREW MEMBER: I tried.

01:46:03 CAPTAIN: Why burn yourself out?

01:46:05 RELIEF: Because I'm a pilot. I made three [trips]. I'm sick and taking medication.

01:48:03 CAPTAIN: Excuse me, Jimmy, while I take a quick trip to the toilet . . . before it gets crowded. While they are eating, and I'll be back to you.

01:48:18.55 [Sound similar to cockpit door operating]

01:48:39.92 RELIEF FIRST OFFICER [WHO PUTS THE AIRCRAFT INTO A 40-DEGREE NOSE-DOWN DIVE]: I rely on God. [Heard faintly]

01:48:57.93 [Series of thumps and clicks starts and continues for approximately seventeen seconds]

01:49:47.54 [Sound of two clicks and two thumps]

01:49:48.42 RELIEF FIRST OFFICER: I rely on God.
01:49:53.32 [One loud thump and three faint thumps]
01:49:57.33 RELIEF FIRST OFFICER: I rely on God.
I rely on God.
01:49:58.78 [Four tones similar to master caution aural beeper]
01:50:00.15 RELIEF FIRST OFFICER: I rely on God.
I rely on God. I rely on God. I rely on God.
01:50:04.72 [Sound of loud thump]
01:50:05.89 RELIEF FIRST OFFICER: I rely on God.
01:50:06.37 CAPTAIN [RETURNS TO COCKPIT]:
What's happening? What's happening?

The accident aeroplane's movements after the command captain returned to the cockpit were the result of both pilot inputs, including opposing elevator inputs as the relief first officer continued to command nose-down and the captain commanded nose-up. Nose-up elevator movements began only after the captain returned to the cockpit. Recovery of the aircraft was possible but it was not accomplished. Seconds after the nose-up elevator movements began, the elevator surfaces began moving in different directions, with the captain's control column commanding nose-up movement and the relief first officer's control column commanding nose-down movement. After the elevator split began, the relief first officer shut down the engines. The captain repeatedly asked him to 'pull with me', but the relief first officer continued to command nose-down elevator movement. The captain's actions were consistent with an attempt to recover the aeroplane while the relief first officer's were not.

117

01:50:07.07 RELIEF FIRST OFFICER: I rely on God.
01:50:07.11 [Sound of numerous thumps and clinks continue for approximately fifteen seconds]
01:50:08.20 [Repeating hi-low tone similar to master caution aural starts and continues to the end of recording]
01:50:08.48 RELIEF FIRST OFFICER: I rely on God.
01:50:08.53 CAPTAIN: What's happening? What's happening, Gamil? What's happening?
01:50:19.51 [Four tones similar to master caution aural beeper]
01:50:24.92 CAPTAIN: What is this? What is this? Did you shut [down] the engine[s]? Get away in the engines. Shut the engines. Pull [back on yoke to pull out of dive]. Pull with me. Pull with me. Pull with me.
01:50:38.47 END OF RECORDING

The NTSB determined that the probable cause of EgyptAir Flight 990's accident was the aircraft's departure from normal cruise flight and subsequent crash into the Atlantic Ocean as a result of the relief first officer's flight 'control inputs'. The reasons for his actions were not determined, but it was widely, and unofficially, assumed that he was committing suicide.

The Egyptian government, which had requested the assistance of the NTSB in the investigation, was not satisfied with its draft conclusions. In a rebuttal, the Egyptian government blamed the accident on the functioning of the aircraft rather than on a pilot losing his mind. In a memorandum to the NTSB, the Egyptian government wrote:

'The NTSB's draft report, prepared under a delegation from the Government of Egypt to the Government of the United States, reflects a limited and incomplete investigation and a corresponding inadequate analysis. More importantly, an objective consideration of the evidence accumulated during the investigation shows that the NTSB, in its draft report, used selected facts and speculative conclusions to support a predetermined theory, instead of determining what probable cause, if any, an unbiased evaluation of all of the evidence would support. In particular, the NTSB's conclusion that the probable cause of the accident is the deliberate action of the relief First Officer is not supported by any evidence of intent or motive that would explain the First Officer's alleged conduct. Indeed, the NTSB omits any discussion of motive and intent and of the facts in the record that squarely contradict a theory of deliberate pilot action. Equally, if not more disturbing, is the NTSB's total disregard of the relevance of the unequivocal evidence of either sheared or deformed bellcrank rivets, not only on EgyptAir 990, but also on other Boeing 767 aircraft. Notably, neither the NTSB nor Boeing is able to explain the cause of these documented abnormalities. These sheared and deformed rivets are direct evidence of a potential defect in the aeroplane's elevator system . . . It is inconceivable that the NTSB can credibly proffer a probable cause based on pilot action when it admits the existence of unexplained damage to a critical elevator component. These shortcomings in both investigation and analysis compelled the

119

Egyptian Civil Aviation Authority (ECAA) to prepare its own report and analysis.'

The ECAA inferred that the NTSB was trying to protect the manufacturer of the accident aeroplane, which was a Boeing 767–366ER.

The NTSB stood by its findings.

MARSA EL-BREGA, LIBYA
13 January 2000

The Shorts 360, a twin-propeller aircraft, leased to Sirte Oil Co. in Libya to transport workers between its headquarters and various oil fields, departed Tripoli at 12.29 en route to Marsa el-Brega. The crew noted a fuel imbalance and did a cross feed until fuel was balanced again at 2.17 p.m. The descent from 7000 feet was started at 2.25. Eleven minutes later both engines flamed out. The crew ditched the aircraft in the water off the Libyan Coast in a 10-degree nose-up attitude. The tail broke off and the wreckage sank to a depth of approximately 125 feet, about three miles from the coast. The captain and first officer both survived and were so busy talking about how to fly a different aircraft from the one they were flying, the Fokker F-28, that they failed to switch on the anti-icing system for the engines as weather conditions deteriorated. As the aircraft came in to land, ice that had built up melted, flooding the engines and cutting off the power. A device that might have restarted the engines was not switched on. There were no lifejackets on board and many of the dead – who accounted for 22 out of the 44 occupants of the aircraft – drowned because they were unaware that their seat cushions doubled as flotation devices and possibly could have saved their lives.

14:35:41 CAPTAIN: Approach checks, yes.

14:35:43 FIRST OFFICER: Air conditioning okay are off. Okay seat belt sign.

14:35:50 CAPTAIN: On.

14:35:51 FIRST OFFICER: Landing lights.

14:35:51 CAPTAIN: Both on.

14:35:54 FIRST OFFICER: Weather radar.

14:35:55 CAPTAIN: On stand-by mode.

14:36:01 FIRST OFFICER: Cabin report.

14:36:03 CAPTAIN: Next.

14:36:22 CONTROL: Go ahead.

14:36:24 FIRST OFFICER: Okay, wind check, please.

14:36:29 CONTROL: 090 variable 120 at 20 knots.

14:36:32 FIRST OFFICER: Shukrn [a phonetic of Arabic].

14:36:57 CAPTAIN: Ah, ah.

14:36:57 FIRST OFFICER: Shino hada [Arabic phonetic].

14:36:58 [Sound of autopilot disconnect]

14:36:58 CAPTAIN: We just had an engine failure.

14:37:00 FIRST OFFICER: You are not kidding.

14:37:01 FIRST OFFICER: Oil pressure going low.

14:37:03 CAPTAIN: Power.

14:37:04 FIRST OFFICER: Okay.

14:37:05 CAPTAIN: Okay, power on the right engine.

14:37:11 FIRST OFFICER: Okay, checked.

14:37:12 [Warning sound]

14:37:21 CAPTAIN: Okay, gears and flaps are up.

14:37:23 FIRST OFFICER: Yes.

14:37:25 CAPTAIN: Confirm left engine failure.

14:37:26 FIRST OFFICER: Confirmed.

14:37:27 CAPTAIN: Shut down left engine.

14:37:28 FIRST OFFICER: Okay.

14:37:28 [Sound of engine running down]

14:37:29 CAPTAIN: Oh, oh, my God.

14:37:33 FIRST OFFICER: What happened?

14:37:34 FIRST OFFICER: Right generator.

14:37:35 CAPTAIN: Both failures, two engine failures.

14:37:38 FIRST OFFICER: Yes.

14:37:39 CAPTAIN: Just call call Marsa Brega.

14:37:40 FIRST OFFICER: Yah, S-21 AAM.

14:37:45 CAPTAIN: Dalila [cabin attendant], prepare for emergency landing. We have two engine failures.

14.37:46 [Warning sound]

14:37:48 FIRST OFFICER: Okay, we lost engine [ya] abdo ssalam [Arabic phonetic].

14:37:50 CAPTAIN: Two engines.

14:37:52 FIRST OFFICER: Two engines, we lost two engines [wanahna] [Arabic phonetic]. Approaching the coast line.

14:37:56 CAPTAIN: My God.

14:37:57 CAPTAIN: Try to restart.

14:37:57 FIRST OFFICER: Okay.

14:37:58 CAPTAIN: Try to restart

14:38:00 [Several warning sounds, ground proximity warning system, for twenty-two seconds]

14:38:10 CAPTAIN: Oh, my God.

14:38:17 CAPTAIN: Open the emergency exit hateh [ya] bashfr [Arabic phonetic].

14:38:19 FIRST OFFICER: Okay [hadi] [Arabic phonetic]?

14:38:21 CAPTAIN: Yes.

14:38:22 FIRST OFFICER: [Ya] basher [Arabic phonetic].

14:38:23 [Sound of opening of emergency hatch]

14:38:25 FIRST OFFICER: [Aiwa] [Arabic phonetic].

14:38:26 FIRST OFFICER: Okay, we are ditching [fi albahr] ya [Arabic phonetic].

14:38:32 CONTROL: Confirm emergency.

14.38:34 FIRST OFFICER: Emergency [fi albahr] [Arabic phonetic].

14:38:34 SOUND OF IMPACT AND END OF RECORDING

OFF ANACAPA ISLAND, California, USA

31 January 2000

At about 2.30 p.m. Alaska Airlines Flight 261, a McDonnell Douglas MD-83, registered as N963AS, took off from Puerto Vallarta, Mexico, bound for Seattle, Washington, with an intermediate stop planned at San Francisco, with two pilots, three cabin crew members, and eighty-three passengers on board. At 4.11 Los Angeles air traffic control asked the condition of the flight and was told that they were troubleshooting a jammed stabilizer. The crew was granted a 20–25,000-foot block altitude clearance. At 4.15 the crew was handed off to Los Angeles sector control. The Alaska Airlines crew reported problems maintaining their altitude; they reported their intention to divert to Los Angeles International Airport. The crew then requested permission to descend to 10,000 feet over water to change their aircraft configuration. Los Angeles cleared them to 17,000 feet. The last message from Flight 261 was a request for another block altitude, which was granted at 4.17. During the descent the crew was also in contact with Alaska Airlines' maintenance base in Seattle and Los Angeles to troubleshoot their stabilizer trim problems. As the crew attempted to diagnose or correct the problem the out-of-trim condition got worse. The aeroplane pitched nose-down. When preparing the plane for landing in Los

Angeles, control was lost suddenly and witnesses on the ground saw the MD-83 'tumbling, spinning, nose down, continuous roll, corkscrewing and inverted'.

15:49:50 ALASKA AIRLINES MAINTENANCE FACILITY: Um, beyond that I have verified no history on your aircraft in the past thirty days.

15:49:57.7 CAPTAIN: Yeah, we didn't see anything in the logbook.

FIRST OFFICER: Why don't you pull your seat forward and I'll just check this pedestal back there.
I don't think there's anything beyond that we haven't checked.

CAPTAIN: See when he's saying pedestal . . . I believe he's talking about this switch that's on the . . . that's on the pedestal.

FIRST OFFICER: Yeah, okay.

CAPTAIN: Do you see anything back there?

FIRST OFFICER: Uh, there's . . .

15:50:40 MAINTENANCE: And 261, maintenance.

15:50:42.0 FIRST OFFICER: Go ahead maintenance, 261.

15:50:44 MAINTENANCE: Understand you're requesting, uh, diversion to LA for this, uh, discrepancy. Is there a specific reason you prefer LA over San Francisco?

15:50:54.4 CAPTAIN: Well, a lotta times it's windy and rainy and wet in San Francisco. And, uh, it seemed to me that a dry runway . . . where the wind is usually right down the runway, seemed a little more reasonable.

126

15:51:09.9 MAINTENANCE: Okay, and, uh . . . is this added fuel that you're gonna have in LA. Gonna be a complication or an advantage?

15:51:18.1 CAPTAIN: Well, the way I'm reading it, uh, heavier aeroplanes land faster . . . Right now I got fifteen five [of fuel] on board. I'm thinking to land with about twelve [12,000 pounds] which is still, uh, an hour and forty minutes . . . uh, and those are the numbers I'm running up here.

15:51:20.6 FIRST OFFICER: LA, Alaska 261, three one zero.

15:51:36 MAINTENANCE: Okay, uh, 261, stand by for dispatch.

15:51:38 FIRST OFFICER: Los Angeles, Alaska 261, three one zero.

15:51:40 CAPTAIN: Okay, [and] the other thing you gotta know is that they're talking about holding and delays in San Francisco, um, for your maintenance facil– eh, you know planning, uh, it, uh, LA seemed like a smarter move from airworthy move.

15:51:42 LOS ANGELES CONTROL: Alaska 261, LA Centre, roger.

15:51:50 FIRST OFFICER: . . .There's two people on the frequency. I'm sorry, Alaska 261, I didn't hear your response.

15:51:58 LOS ANGELES CONTROL: Alaska 261 squawk two zero one zero.

15:52:01 FIRST OFFICER: Two zero one zero, Alaska 261.

15:52:02 SEATTLE DISPATCH: 261, dispatch . . . uh, current San Francisco weather one eight zero at six, nine

127

miles, few at fifteen hundred broken twenty-eight hundred overcast thirty-four hundred . . . uh, if, uh, you want to land at LA of course for safety reasons we will do that, uh, we'll, uh, tell you though that if we land in LA, uh, we'll be looking at probably an hour to an hour and a half. We have a major flow programme going right now. Uh, that's for ATC back in San Francisco.

15:52:31 CAPTAIN: Well, uh, boy, you put me in a spot here. I really didn't want to hear about the flow being the reason you're calling us 'cause I'm concerned about overflying suitable airports.

15:52:51 SEATTLE DISPATCH: Well, we wanna do what's safe, so if that's what you feel is safe we just wanna make sure you have all of the info.

15:52:59 CAPTAIN: Yeah, we kinda assumed that we had . . . what's the, uh, the wind again there in San Francisco?

15:53:03 SEATTLE DISPATCH: Wind at San Francisco currently zero uh one zero eight at six.

15:53:08 FIRST OFFICER: What runway they landing . . . 10?

15:53:09 CAPTAIN: What's that?

15:53:10 FIRST OFFICER: Ask him what runway they're landing.

15:53:11 CAPTAIN: And confirm they're landing runway 10?

15:53:15 DISPATCH: And, uh, stand by. I'll confirm that.

15:53:17 FIRST OFFICER: And see if the runways are dry or wet.

15:53:19 CAPTAIN: And we need to know if they're dry or wet.

15:53:21 DISPATCH: Yup, I'll, uh, find that out and, uh, correction on that wind [at] one eight zero at six, and stand by.

15:53:28 CAPTAIN: One eight zero at six . . . so that's runway 16 what we need is runway 19, and they're not landing runway 19.

15:53:35 FIRST OFFICER: I don't think so. We might just ask if there's a ground school instructor there available and discuss it with him . . . or, uh, simulator instructor.

15:53:40 CAPTAIN: Yes. [TO DISPATCH] And, uh, dispatch, 261 . . . we're wondering if we can get some support out of the instructional force – instructors up there, if they got any ideas on us. [CAPTAIN TO FIRST OFFICER] You're talkin' to ATC, huh?

15:54:24 FIRST OFFICER: Yeah, uh huh. Let's confirm the route of flight. It's, uh, I wasn't totally sure but it's, uh, direct oceanside?

15:54:32 CAPTAIN: Tijuana oceanside . . . oceanside right . . . then Santa Catalina. Somebody was callin' in about wheelchairs when I'm workin' a problem.

15:54:50 FLIGHT ATTENDANT: Oh really?

15:54:50 CAPTAIN: Okay, yeah, now . . . That's something that oughta be in the computers. If they want it that bad you guys oughta be able to pick up the phone – just . . . drives me nuts. Not that I wanna go on about it . . . you know. It just blows me away. They think we're gonna land, they're gonna fix it, now they're worried about the flow. I'm sorry [but] this aeroplane isn't gonna go anywhere for a while . . . so you know.

15:55:16 FLIGHT ATTENDANT: So they're trying to put the pressure on you –

15:55:18 CAPTAIN: Well, no, yeah.

15.55:19 FLIGHT ATTENDANT: Well, get it to where it needs to be.

15:55:20 CAPTAIN: And actually it doesn't matter that much to us.

15:55:23 FLIGHT ATTENDANT: Still not gonna go out on time to the next . . .

15:55:24 CAPTAIN: Yeah . . . I thought they'd cover the people better from LA, [than] San Francisco.

15:55:32 FIRST OFFICER: LA, Alaska 261, just confirm our routing after Tijuana is, direct oceanside?

15:55:38 LOS ANGELES CONTROL: Alaska 261, after Tijuana cleared to San Francisco via direct San Marcos . . . maintain flight level three one zero [31,000 feet].

15:55:47 FIRST OFFICER: Okay, San Francisco. San Marcos . . . direct three one zero, Alaska 261.

15:56:03 SEATTLE DISPATCH: Alaska 261, dispatch.

15:56:06 CAPTAIN: Dispatch, Alaska 261, go ahead.

15:56:08 DISPATCH: I called ATIS [Los Angeles Airport Automated Terminal Information System]. They're landing twenty-eight right. Twenty-eight left wasn't able to get the runway report but looking at past weather it hasn't rained there in hours so I'm looking at probably a dry runway.

15:56:21 CAPTAIN: Okay, uh. I have the information I have available to me and we're waitin' on that update. I'm looking at an approach speed of a hundred and eighty knots. Do you have a wind at LAX [Los Angeles Airport]?

15:56:50 DISPATCH: It's two six zero at nine.

15:56:56 CAPTAIN: Okay, two six at nine . . . versus a direct crosswind, which is effectively no change in groundspeed . . . I gotta tell you, when I look at it from a safety point, I think that something that lowers my groundspeed makes sense.

15:57:16 DISPATCH: Okay, 261, that'll mean LAX then for you. I was gonna get you, if I could to call LAX with that info and they can probably whip out that captain for you real quick.

15:57:30 CAPTAIN: I suspect that's what we'll have to do. Okay, here's my plan. We're gonna continue as if going to San Francisco, get all that data, then begin our descent back into LAX, and at a lower altitude. We will configure and check the handling envelope before we proceed with the approach.

15:58:05 DISPATCH: Okay, 261, dispatch copied that. If you can now keep LA ops updated on your ETA, that would be great, and I'll be talking with them.

15:58:15 CAPTAIN: Okay, well if you'll let them know we're comin' here I think they'll probably listen as we talk . . . we're goin to LAX. We're gonna stay up here and burn a little more gas, get all our ducks in a row, and then we'll be talking to LAX when we start down to go in there.

15:58:29 DISPATCH: Okay, and if you have any problems with them giving you a CG [centre of gravity], gimme a call back.

15:58:34 CAPTAIN: Okay, break. LAX do you read Alaska 261?

15:58:39 LOS ANGELES OPERATIONS: 261, I do copy. Do you have an ETA for me?

15:58:45 CAPTAIN: I'm gonna put it at about thirty, thirty-five minutes . . . The longer the more fuel I burn off the better I am . . . but I wonder if you can compute our current CG based on the information we had at takeoff for me.

15:58:58 OPERATIONS: Okay, your transmission is coming in broken but go ahead.

15:59:02 CAPTAIN: You know what? I'll wait a minute. We'll be a little bit closer and that'll help everything.

15:59:06 OPERATIONS: Okay, also 261, just be advised because you're an international arrival we have to get landing rights. I don't know how long that's gonna take me . . . but I have to clear it all through customs first.

15:59:19 CAPTAIN: Okay, I understand . . . I remember this is complicated. Yeah, well, better start that now 'cause we are comin' to you. [TO FIRST OFFICER] We'll call 'em back as we get closer to Catalina.

15:59:34 FIRST OFFICER: As we get what?

15.59:34 CAPTAIN: Closer to LA. She's got to get landing rights.

15:59:37 FIRST OFFICER: We're ninety-four miles from LA now.

15:59:38 CAPTAIN: Oh, okay . . . You wanna listen to the ATIS?

15:59:42 FIRST OFFICER: In fact I switched it once already, just kinda late.

15:59:44 CAPTAIN: You got the jet.

15:59:44 FIRST OFFICER: I got it.

15:59:50 ATIS [RECORDED]: Charlie five and Charlie

six is restricted . . . taxiway Charlie five is restricted to MD eleven and smaller. Read back all runway. Hold short instructions. Upon receipt of your ATC clearance read back only your call sign and transponder code unless you have a question. Advise on initial contact [that] you have information mike. Los Angeles International Airport information mike. Two two five zero Zulu. Wind two three zero at eight. Visibility eight. Few clouds at two thousand eight hundred. One two thousand scattered. Ceiling two zero thousand overcast. Temperature one six dew point one one. Altimeter three zero one seven. Simultaneous ILS approaches in progress runway twenty-four right and twenty-five left or vector for visual approach will be provided. Simultaneous visual approaches to all runways are in progress. And parallel localizer approaches are in progress between Los Angeles International and Hawthorne airports. Simultaneous instrument departure in progress runway two four and two five. Notices to airmen.

16:01:01 FIRST OFFICER: So he wanted us to go to San Fran initially?

16:01:06 CAPTAIN: To keep the schedule alive. I mean it was just . . . I mean, he had all the reasons to do it. I stated concern about . . . overflying a suitable airport –

16:01:15 FIRST OFFICER: Yeah.

16:01:16 CAPTAIN: – but I was listening, then when he gives me the wind, it's . . . the wind was a ninety degree cross at ten knots. Two eight and we'd be landing on – you know I don't know . . . I wrote it down there . . . the winds were . . . one eighty at six . . . I don't know. I don't care . . . you know what? I expect him to figure all that shit –

16:01:53 FIRST OFFICER: Right.

16:01:53 CAPTAIN: He's got it on the screen –

16:01:54 FIRST OFFICER: That's why I was thinking that an instructor would really cut through the crap there. They . . . not available?

16:02:00 CAPTAIN: Well, they just don't talk to each other. I mean I . . .

16:02:04 FIRST OFFICER: They've always told us they were available you know anytime you have a problem.

16:02:12.6 CAPTAIN: Los Angeles, 261, do you read me better now?

16:02:31 LOS ANGELES OPERATIONS: Go ahead 261.

16:02:33.6 CAPTAIN: 261, I know you're busy on us, but we're discussing it up here. Could you give us the winds at San Francisco. If you could just pull 'em up on your screen?

16:03:00 OPERATIONS: Okay, ahhh San Francisco, okay, we've got uh . . . winds are one seventy at six knots.

16.03:15.6 CAPTAIN: Okay, thank you. That's what I needed. We are comin' in to see you . . . And I've misplaced the paperwork here.

16:03:23 FIRST OFFICER: There it is.

16:03:35 CAPTAIN: I can't read your writing . . . Can you read her the zero fuel weight – and all those numbers.

16:03:40 FIRST OFFICER: Yeah.

FIRST OFFICER TO OPERATIONS: Uh, 261 . . . do you need our, uh, our numbers?

16:.03:52 OPERATIONS: Yeah, we just wanna advise that we do not have landing rights as yet.

16:03:56 FIRST OFFICER: Here's our numbers. We had ten in first class, seventy in coach, zero fuel weight one zero two one one zero point one, fuel on board, thirty-four point niner takeoff weight one thirty-six five one one point eight, CG eleven point eight.

16:04:19 OPERATIONS: Okay, I got ten and seventy Z fuel weight one zero two one one zero point one, fuel on board thirty-four decimal nine takeoff weight five one one decimal eight and a CG of eleven decimal eight.

16:04:32 FIRST OFFICER: Yeah, uh, takeoff one three six five one one point eight and uh, CG one one point eight. And we currently have thirteen thousand six hundred pounds of fuel on board.

16:04:43 CAPTAIN: Estimate ten thousand on landing.

16:04:45 FIRST OFFICER: Estimating ten thousand pounds on landing.

16:04:53 OPERATIONS: Okay, you said your takeoff weight was . . . one one, one five one one decimal eight?

16:04:58 FIRST OFFICER: One three six five one one point eight.

16.05:05 OPERATIONS: One three six five one one point eight, thank you.

16:05:07 FIRST OFFICER: And we're currently a hundred and fifteen seven on our weight, and we'll burn another three thousand pounds. So our . . . actually our landing speed will be one forty-eight plus . . . some additive right?

16:07:06 CAPTAIN: Let's guess . . . let's guess one twelve.

16:07:10 FIRST OFFICER: Okay.

16:07:10 CAPTAIN: One forty six . . . plus . . . I get a

minus two, worst case . . . twenty-four knots . . . fifty sixty seventy . . .

16:07:33 OPERATIONS: Alaska 261, from operations, can you give us your tail number?

16:07:38 CAPTAIN: Uh, two sixty-one, it was ship number nine six three.

16:07:43 OPERATIONS: Copy that two . . . uh, your aircraft number is nine six three.

16:07:47 CAPTAIN: Affirmative, thank you.

16:07:51 MAINTENANCE FACILITY LOS ANGELES: And 261 maintenance.

16:07:53 CAPTAIN: 261, go.

16:07:54 MAINTENANCE: Yeah, are you guys with the, uh, horizontal situation?

16:07:58 CAPTAIN: Affirmative.

16:07:59 MAINTENANCE: Did you try the suitcase handles and the pickle switches . . . ?

16:08:03 CAPTAIN: We tried everything together, uh . . . We've run just about everything. If you've got any hidden circuit breakers we'd love to know about 'em.

16:08:14 MAINTENANCE: I'm off. I'll look at the circuit breaker guide just as a double check and, yeah, I just wanted to know if you tried the pickle switches and the suitcase handles to see if it was movin' in with any of the other switches other than the suitcase handles alone or nothing.

16:08:29.9 CAPTAIN: We tried just about every iteration.

16:08:32 MAINTENANCE: And alternate's inop, too, huh?

16:08:35.1 CAPTAIN: Yup, it just appears to be

jammed the whole thing. It spikes out when we use the primary. We get AC load that tells me the motor's tryin' to run but the brake won't move it. When we use the alternate, nothing happens.

16:08:50 MAINTENANCE: You you you say you get a spike when on the meter there in the cockpit, when you, uh, try to move it with the, uh, um, with the primary right?

16:08:59 CAPTAIN: I'm gonna click it off. You got it.

16:09:00 FIRST OFFICER: Okay.

16:09:01.5 CAPTAIN: Affirmative. We get a spike when we do the primary trim but there's no appreciable change in the electrical when we do the alternate.

16:09:09 MAINTENANCE: Thank you, sir, see you here.

16:09:13 CAPTAIN: Let's do that.

16:09:14.8 [Sound of click]

16:09:14.8 CAPTAIN: This'll click it off.

16.09:16 [Sound of clunk]

16:09:16.9 [Sound of two faint thumps in short succession]

16:09:17.0 [Sound similar to horizontal stabilizer-in-motion audible tone]

16:09:18 CAPTAIN: Holy shit.

16:09:19.6 [Sound similar to horizontal stabilizer-in-motion audible tone]

16.09:21 CAPTAIN TO FIRST OFFICER: You got it? . . . Fuck me.

16:09:24 FIRST OFFICER: What are you doin'?

16:09:25 CAPTAIN: It clicked off –

16.09:25.4 [Sound of chime] *Altitude*.

16:09:26 CAPTAIN: – it . . . got worse . . . okay.

16:09:30 [Sound similar to airframe vibration begins]

16:09:31 CAPTAIN: You're stalled.

16:09:32 [Sound similar to airframe vibration becomes louder]

16:09:33 CAPTAIN: No, no, you gotta release it. Ya gotta release it.

16:09:34 [Sound similar to airframe vibration ends]

16:09:42.4 CAPTAIN: Let's . . . speed brake. Gimme a high pressure pumps.

16:09:52 FIRST OFFICER: Okay.

16:09:52 CAPTAIN: Help me back, help me back. Centre, Alaska 261, we are in a dive here. And I've lost control, vertical pitch.

16:10:01.9 *Overspeed.* [Begins and repeats for approximately thirty-three seconds]

16:10:05 LOS ANGELES CONTROL: Alaska 261, say again, sir.

16:10:06.6 CAPTAIN: We're out of twenty-six thousand feet, [and] we are in a vertical dive.

16:10:20 CAPTAIN TO FIRST OFFICER: Just help me. Once we get the speed slowed maybe . . . we'll be okay. We're at twenty-three thousand seven hundred . . . request, uh. Yeah, we got it back under control here.

16:10:34 FIRST OFFICER: No, we don't, okay.

16:10:37 LOS ANGELES CONTROL: The altitude you'd like to remain at?

16:10:45 FIRST OFFICER: Let's take the speedbrakes off . . . I'm –

16:10:46 CAPTAIN: No, no, leave them there. It seems to be helping.

16.10:51 FIRST OFFICER: Fuck me.

16:.10:53 [Sound of chime] *Altitude.*

16:10:55 CAPTAIN: It really wants to pitch down.

16:10:58 FIRST OFFICER: Okay.

16:10:59 CAPTAIN: Don't mess with that.

16:11:04 FIRST OFFICER: I agree with you.

16:11:04 CONTROL: Alaska 261, say your condition.

16:11:06.6 CAPTAIN: 261, we are at twenty-four thousand feet, kinda stabilized. We're slowing here, and, uh, we're gonna, uh, do a little troubleshooting. Can you gimme a block between uh, twenty and twenty-five [thousand feet]?

16:11:21 CONTROL: Alaska 261, maintain block altitude flight level two zero zero through flight level two five zero.

16:11:27 CAPTAIN: Alaska 261, we'll take that block. We'll be monitor'n' the freq.

16:11:31 FIRST OFFICER: You have the aeroplane. Let me just try it.

16:11:33 CAPTAIN: Okay.

16.11.33 FIRST OFFICER: Uh, how hard is it?

16:11:33 CAPTAIN: I don't know. My adrenaline's goin' . . . It was really tough there for a while.

16:11:38 FIRST OFFICER: Yeah, it is.

16:11:39 CAPTAIN: Okay.

16:11:43 FIRST OFFICER: Whatever we did is no good. Don't do that again.

16:11:44 CAPTAIN: Yeah, no, it went down, it went to full nose down.

16:11:48 FIRST OFFICER: It's a lot worse than it was?

16:11:50 CAPTAIN: Yeah, yeah, we're in much worse

shape now. I think it's at the stop, full stop . . . And I'm thinking, we can [if] it go any worse . . . but it probably can . . . but when we slowed down, let's slow it, let's get down to two hundred knots and see what happens.

16:12:16 FIRST OFFICER: Okay? We have to put the slats out and everything . . . flaps and slats.

16:12:20 CAPTAIN: Well, we'll wait. Okay. You got it for a second?

16:12:23 FIRST OFFICER: Yeah.

16:12:25.3 CAPTAIN: Maintenance 261, are you on?

16:12:30 MAINTENANCE: Yeah, 261, this is maintenance.

16:12:32.0 CAPTAIN: Okay, we did – we did both the pickle switch and the suitcase handles and it ran away full nose trim down.

16:12:39 MAINTENANCE: It ran away trim down?

16:12:42 CAPTAIN: And now we're in a . . . pinch so we're holding . . . we're worse than we were.

16:12:50 MAINTENANCE: Okay, uh . . . geeze. You want me to talk to 'em? 261, maintenance, you getting full nose trim down but are you getting any you don't get no nose trim up, is that correct?

16:13:04 CAPTAIN: That's affirm. We went to full nose down and I'm afraid to try it again to see if we can get it to go in the other direction.

16:13:10 MAINTENANCE: Okay, well your discretion. If you want to try it, that's okay with me. If not that's fine. Um, we'll see you at the gate.

16:13:20 FIRST OFFICER: Did it happen . . . when you pulled back? It went forward?

16.13:22 CAPTAIN: I went tab down . . . right, and it

should have come back. Instead it went the other way. What do you think? You wanna try it or not?

16:13:32 FIRST OFFICER: Uhh, no. Boy, I don't know.

16:13:33 CAPTAIN: It's up to you, man.

16:13:35 FIRST OFFICER: Let's head back towards . . . Here let's see . . . well, we're –

16:13:39 CAPTAIN: I like where we're goin' out over the water myself . . . I don't like goin' this fast though. Okay, you got . . . second?

16:13:58 FIRST OFFICER: Yeah. We better . . . talk to the people in the back there [passengers].

16:14:03 CAPTAIN: Yeah, I know.

16:14:04 LOS ANGELES CONTROL: Alaska 261, let me know if you need anything.

16:14:08 FIRST OFFICER: Yeah, we're still workin' this.

16:14:12 PUBLIC ADDRESS/CAPTAIN: Folks, we have had a flight control problem up front here. We're workin' it. That's Los Angeles off to the right there. That's where we're intending to go. We're pretty busy up here workin' this situation. I don't anticipate any big problems once we get a couple of subsystems on the line. But we will be going into LAX and I'd anticipate us parking there in about twenty to thirty minutes.

16:14:39 CAPTAIN TO FIRST OFFICER: Okay . . . did the, first of all, speedbrakes. Did they have any effect? Let's put the power where it'll be for one point two, for landing. You buy that? Slow it down and see what happens.

16:14:54 CONTROL: Alaska 261, contact LA Centre

one two six point five two. They are aware of your situation.

16:15:00.0 FIRST OFFICER: Alaska 261, say again. The frequency, one two zero five two?

16:15:02 CAPTAIN: I got the yoke.

16:15:04 CONTROL: Alaska 261, twenty-six fifty-two.

16:15:06 FIRST OFFICER: Thank you.

16:15:07 LOS ANGELES CONTROL: You're welcome. Have a good day.

16:15:19.7 FIRST OFFICER: LA, Alaska 261, we're with you. We're at twenty-two five. We have a jammed stabilizer and we're maintaining altitude with difficulty. Uh, but uh, we can maintain altitude we think . . . and our intention is to land at Los Angeles.

16:15:36 CONTROL: Alaska 261, LA Centre, roger. Um, you're cleared to Los Angeles airport via present position direct, uh, Santa Monica, direct Los Angeles and, uh, you want lower now or what do you want to do, sir?

16:15:54 CAPTAIN: Let me get, let me have it. Centre, uh, Alaska 261. I need to get down about ten, change my configuration, make sure I can control the jet and I'd like to do that out here over the bay if I may.

16:16:07 CONTROL: Alaska 261, roger that. Stand by here.

16:16:11 FIRST OFFICER: Let's do it at this altitude instead –

16:16:11 CAPTAIN: What?

16:16:12 FIRST OFFICER: – of goin' to ten. Let's do it at this altitude . . .

16:16:14 CAPTAIN: . . . 'cause the airflow's that much

difference down at ten. This air's thin enough that you know what I'm sayin'?

16:16:20 FIRST OFFICER: Yeah, uh, I'll tell 'em to, uh –

16:16:22 CAPTAIN: I just made a PA to everyone to get everybody – down. You might call the flight attendants.

16:16:27 [Sound similar to cockpit door operating]

16:16:32 FLIGHT ATTENDANT: I was just comin' up this way.

16:16:32 CONTROL: Alaska 261, fly a heading of two eight zero and descend and maintain one seven thousand.

16:16:39.0 CAPTAIN: Two eight zero and one seven seventeen thousand Alaska 261. And we generally need a block altitude.

16:16:45 CONTROL: Okay, and just, um, I tell you what. Do that for now, sir, and contact LA Centre on one three five point five. They'll have further instructions for you, sir.

16:16:56.9 FIRST OFFICER: Thirty-five five, say the altimeter setting?

16:16:59 CONTROL: The LA altimeter is three zero one eight.

16:17:01 CAPTAIN TO FLIGHT ATTENDANT: I need everything picked up – and everybody strapped down –

16:17:04 FLIGHT ATTENDANT: Okay.

16:17:04 CAPTAIN: – 'cause I'm gonna unload the aeroplane and see if we can – we can regain control of it that way.

16:17:09 FLIGHT ATTENDANT: Okay, we had like a big bang back there –

16:17:11 CAPTAIN: I heard it – the stab trim I think it –

16:17:13 FIRST OFFICER: You heard it in the back?

16:17:13 FLIGHT ATTENDANT: Yeah.

16:17:15 CAPTAIN: I think the stab trim thing is broke –

16:17:17 FLIGHT ATTENDANT: – I didn't wanna call you guys . . . but – they're like, you better go up there –

16:17:21 CAPTAIN: I need you [to get] everybody strapped in now, dear. 'Cause I'm going to release the back pressure and see if I can get it . . . back.

16:17:30 [Sound similar to cockpit door operating]

16:17:33 FIRST OFFICER: Three zero one eight.

16:17:37 CAPTAIN: I'll get it here.

16:17:40 FIRST OFFICER: I don't think you want any more speedbrakes, do you?

16:17:42 CAPTAIN: Uh, no, actually.

16:17:46 FIRST OFFICER: He wants us to maintain seventeen [thousand feet].

16:17:51 CAPTAIN: Okay, I need help with this here. Slats ext – let's –

16:17:54 FIRST OFFICER: Okay, slats –

16:17:54 CAPTAIN: Gimme slats extend.

16:17:55 FIRST OFFICER: Got it.

16:17:56.6 [Sound similar to slat/flap handle movement]

16:17:58 CAPTAIN: I'm test flyin' now –

16.17:59 FIRST OFFICER: How does it feel?

16:18:00 CAPTAIN: It's wantin' to pitch over more on you.

16:18:02 FIRST OFFICER: Really? Try flaps? . . . fifteen, eleven?

16:18:05 CAPTAIN: Ah, let's go to eleven.

16:18:07.3 [Sound similar to slat/flap handle movement]

16:18:09 FIRST OFFICER: Okay . . . get some power on.

16:18:10 CAPTAIN: I'm at two hundred and fifty knots, so I'm lookin' . . .

16:18:17 FIRST OFFICER: Real hard?

16:18:17 CAPTAIN: No, actually it's pretty stable right here . . . See but we got to get down to a hundred an' eighty [knots]. Okay . . . bring the flaps and slats back up for me.

16:18:32 FIRST OFFICER: Slats, too?

16:18:33 CAPTAIN: Yeah.

16:18:36.8 [Sound similar to slat/flap handle movement]

16:18:37 FIRST OFFICER: That gives us . . . twelve thousand pounds of fuel. Don't overboost them.

16:18:47 CAPTAIN: What I'm, what I wanna do . . .

16:18:48 [Sound similar to slat/flap handle movement]

16:18:49 CAPTAIN: . . . Get the nose up . . . and then let the nose fall through and see if we can stab it when it's unloaded.

16:18:54 [Sound of chime] *Altitude* [Repeats for some thirty-four seconds]

16:18:56 FIRST OFFICER: You mean use this again? I don't think we should . . . if it can fly, its like –

16:19:01 CAPTAIN: It's on the stop now, it's on the stop.

16:19:04 FIRST OFFICER: Well, not according to that, it's not. The trim might be, and then it might be, uh, if something's popped back there –

16:19:11 CAPTAIN: Yeah.

16:19:11 FIRST OFFICER: It might be . . . mechanical damage, too. I think if it's controllable, we oughta just try to land it –

16:19:16 CAPTAIN: You think so? Okay, let's head for LA.

16:19:21.1 [Sound of faint thump]

16:19:24 FIRST OFFICER: You feel that?

16:19:25 CAPTAIN: Yeah. Okay, gimme sl – see, this is a bitch.

16:19:31 FIRST OFFICER: Is it?

16:19:31 CAPTAIN: Yeah.

16.19:32.8 [Sound of two clicks similar to slat/flap handle movement]

16:19:36.6 [Sound of extremely loud noise] [Increase in background noise begins and continues to end of recording] [Sound similar to loose articles moving around in cockpit]

16:19:43 FIRST OFFICER: Mayday.

16:19:49 CAPTAIN: Push and roll, push and roll. Okay, we are inverted . . . and now we gotta get it . . .

16:19:59 [Sound of chime]

16:20:03 CAPTAIN: Kick. Push push push . . . Push the blue side up. Push.

16:20:14 FIRST OFFICER: I'm pushing.

16:20:16 CAPTAIN: Okay, now let's kick rudder . . . left rudder left rudder.

16:20:18 FIRST OFFICER: I can't reach it.

16:20:20 CAPTAIN: Okay, right rudder . . . right rudder. Are we flyin'? . . . We're flyin' . . . We're flyin' . . . Tell 'em what we're doin'.

16:20:33 FIRST OFFICER: Oh, yeah, let me get . . .

16:20:38 CAPTAIN: Gotta get it over again . . . at least upside down. We're flyin'.
16:20:49 [Sounds similar to compressor stalls begin and continue to end of recording]
16:20:49 [Sound similar to engine spool down]
16.20:54 CAPTAIN: Speedbrakes.
16:20:55.1 FIRST OFFICER: Got it.
16:20:56.2 CAPTAIN: Ah, here we go.
16:20:57.1 END OF RECORDING

The aircraft crashed off Point Mugu in 650 feet of water. All crew members and passengers died.

The NTSB determined that the probable cause of this accident was a loss of aeroplane pitch control resulting from the in-flight failure of the horizontal stabilizer trim system jackscrew assembly's acme nut threads. The thread failure was caused by excessive wear resulting from Alaska Airlines' insufficient lubrication of the jackscrew assembly. Contributing to the accident were Alaska Airlines' extended lubrication interval and the Federal Aviation Administration's approval of that extension, which increased the likelihood that missed or inadequate lubrication would result in excessive wear of the acme nut threads. Also contributing to the accident was the absence on the McDonnell Douglas MD-80 of a fail-safe mechanism to prevent the catastrophic effects of total acme nut thread loss.

RANCHO CORDOVA, California, USA

16 February 2000

At 7.49 p.m. an Emery Worldwide Airlines Flight 17, a McDonnell Douglas DC-8–71F (DC-8), registered as N8079U, took off from Sacramento Airport with two pilots and a flight engineer on board. Earlier that day, Emery had operated the aeroplane on two cargo-carrying flight segments. The aeroplane had arrived at Sacramento at about 6.15 p.m. Cargo handlers unloaded the inbound cargo from the main and belly cargo compartments, and then loaded the outbound cargo, finishing at about 7.30 p.m. Minor maintenance fixes (for example, an inoperative fuel valve indication and a malfunctioning navigation light) needed to be carried out, but neither the flight engineer nor the mechanics observed any significant problems during preflight inspections. The cargo loading supervisor completed weight and balance checks, and the doors were closed for departure. At 19:27:25, the first officer (the flying pilot for the flight) conducted the takeoff briefing. At 19:39:19, the flight engineer advised ground personnel, 'We have four good [engine starts]. You can go ahead and pull the [auxiliary] air and power.'

At 19:48:40, with the flight in position and ready for takeoff, the captain said, 'Airspeed's alive', and the first officer responded, 'Alive here.' The aircraft accelerated down the runway through 80 knots. As the takeoff roll continued, the captain said, 'V one' and 'rotate'. At 19:49:09, the captain said, 'Watch the tail.' At 19:49:44, the flight engineer said, 'We're sinking. We're going down, guys.' Two seconds later, the ground proximity warning system went, '*Whoop, Whoop, Pull up.*' At this time the aeroplane was descending through 679 feet in a steep left bank of about 11 degrees. At 19:49:54 the aircraft climbed through 673 feet and the first officer said, 'Push', and the flight engineer said, 'Okay, so we're going back up.' As the aeroplane's altitude was increasing its left bank also increased, reaching about 45 degrees, and then began to decrease. At 19:49:57, the flight engineer said, 'There you go.' The captain said, 'Roll out' and there was the sound of someone exhaling tensely.

19:42:17 CAPTAIN: Ready on the rudders?
19:42:18 FLIGHT ENGINEER: Yep.
19:42:20 FIRST OFFICER: You're, ah . . . clear right.
19:42:21 CAPTAIN: Left rudder. Centre.
19:42:23 FLIGHT ENGINEER: Checked.
19:42:24 CAPTAIN: Right rudder. Centre.
19:42:25 FLIGHT ENGINER: Checked.
19:42:27 CAPTAIN: Elevator forward. Coming back.
19:42:30 CAPTAIN: Taxi check.
19:42:31 FLIGHT ENGINEER: Taxi checklist. Flaps and slots.
19:42:36 CAPTAIN: Ah, fifteen. Fifteen. Fifteen. Slot light's out.

19:42:39 FIRST OFFICER: Fifteen. Fifteen. Fifteen. Slot light's out.

19:42:43 FLIGHT ENGINEER: Controls EPI.

19:42:44 FIRST OFFICER: Checked.

19:42:45 FLIGHT ENGINEER: Checked. Fuel panel set. Spoilers . . .

19:42:49 FIRST OFFICER: . . . retracted. Lights out.

19:42:51 FLIGHT ENGINEER: Fuel levers . . .

19:42:52 FIRST OFFICER: On in detent.

19:42:53 FLIGHT ENGINEER: Yaw damper . . .

19:42:54 FIRST OFFICER: On.

19:42:55 FLIGHT ENGINEER: Stabilizer . . .

19:42:58 CAPTAIN: One point six.

19:43:00 FIRST OFFICER: Set one point six.

19:43:01 FLIGHT ENGINEER: One six. Shoulder harness.

19:43:06 CAPTAIN: [Clear] on the left.

19:43:06 FIRST OFFICER: [Clear] on the right.

19:43:08 FLIGHT ENGINEER: Take off data briefing.

19:43:13 CAPTAIN: Set left reviewed.

19:43:15 FIRST OFFICER: Set and reviewed.

19:43:19 FLIGHT ENGINEER: Flight nav instruments.

19:43:21 CAPTAIN: Set left.

19:43:22 FIRST OFFICER: Set right.

19:43:23 FLIGHT ENGINEER: Taxi checklist complete.

19:43:43 FIRST OFFICER: [Observing a helicopter on or near airport] That'd be fun. I've never been, I've been in one of those Airstar helicopters, you know, like the Cadillac of helicopters. I've never really been in a helicopter, you know.

19:43:53 FLIGHT ENGINEER: I went up one those R-22 Robinsons.

19:43:56 FIRST OFFICER: Yeah . . .

19:43:56 FLIGHT ENGINEER: That was a . . . thing.

19:43:57 FIRST OFFICER: Now that was a helicopter.

19:43:59 FLIGHT ENGINEER: Went up to [elevation] and did some auto-rotations. That was a blast. [It was] really weird going that slow in the air, though. I don't like it.

19:44:09 FIRST OFFICER: Hey, you're, you're hanging by that bolt, you know [in a helicopter].

19:44:12 FLIGHT ENGINEER: Yeah . . . Jesus nut.

19:46:58 FIRST OFFICER TO CONTROL: Sacramento departure Emery 17 heavy number 122 left Mather. Need our release to ah . . . Dayton.

19:47:14 CONTROL: Emery 17 heavy Sacramento approach, you're released for departure. Report airborne.

19:47:20 FIRST OFFICER: Emery 17 heavy, we'll call you in the air.

19:47:25 [Sound similar to brake release]

19:47:28 CAPTAIN: Before takeoff checklist.

19:47:29 [Sound similar to increasing engine rpm]

19:47:30 FIRST OFFICER TO CONTROL: Mather area traffic, Emery 17 heavy, runway 22 left. [It will] be a left downwind departure Mather.

19:47:40 FIRST OFFICER: Okay, you're clear on the right.

19:47:52 FLIGHT ENGINEER: Anti-skid . . .

19:47:55 FIRST OFFICER: Armed. Light's out.

19:47:56 FLIGHT ENGINEER: Ignition.

19:47:57 FIRST OFFICER: All engines both.

19:47:58 FLIGHT ENGINEER: Transponder. DME.

19:47:59 FIRST OFFICER: On. Reverse switch. Open pressure checks.

19:48:02 FLIGHT ENGINEER: Spoiler pumps normal. Stand by rudder pump start. Packs are off. Boost pumps boost and feed. Landing lights.

19:48:07 CAPTAIN: On.

19:48:08 FLIGHT ENGINEER: Parking brake.

19:48:10 CAPTAIN: Released.

19:48:10 FLIGHT ENGINEER: Before takeoff checklist complete.

19:48:11 CAPTAIN: Your brakes.

19:48:12 FIRST OFFICER: Yeah.

19:48:15 [Sound similar to increasing engine rpm]

19:48:23 FIRST OFFICER: Stand by four. There we go.

19:48:24 FLIGHT ENGINEER: Four spooled.

19:48:40 CAPTAIN: Set reduced. Airspeed's alive.

19:48:44 FIRST OFFICER: Alive here.

19:48:50 CAPTAIN: Eighty knots.

19:48:51 FIRST OFFICER: Eighty knots . . . elevator checks.

19:49:02 CAPTAIN: V one. Rotate.

19:49:08 [Sound similar to stabilizer trim in motion alert]

19:49:09 CAPTAIN: . . . watch the tail.

19:49:11 [Sound similar to stabilizer trim in motion alert]

19:49:12 [Sound similar to stabilizer trim in motion alert]

19:49:13 CAPTAIN: V two.

19:49:14 [Sound similar to stabilizer trim in motion alert]

19:49:14 CAPTAIN: Positive rate.

19:49:16 FIRST OFFICER: I got it.

19:49:17 CAPTAIN: You got it?

19:49:17 FIRST OFFICER: Yep.

19:49:18 CAPTAIN: All right.

19:49:19 FIRST OFFICER: We're going back.

19:49:20 FLIGHT ENGINEER: What the hell?

19:49:20 FIRST OFFICER: Captain's waay out of limits.

19:49:25 FLIGHT ENGINEER: Do you want to pull the power back?

19.49:27 [Sound similar to decreasing engine rpm]

19:49:29 [Sound similar to stick shaker]

19:49:30 FIRST OFFICER: Oh, shit.

19:49:30 CAPTAIN: Push forward.

19:49:31 FIRST OFFICER: Goddd . . . Shit. God.

19:49:36 CAPTAIN TO CONTROL: Emery 17 emergency.

19:49:38 FIRST OFFICER: Ahhh, shit.

19:49:40 CONTROL: Emery 17, Sacramento departure radar contact, say again?

19:49:40 FIRST OFFICER: You steer. I'm pushing.

19:49:44 CAPTAIN TO CONTROL: Emery 17 has an emergency.

19:49:44 FLIGHT ENGINEER: We're sinking. We're going down, guys.

19:49:46 CONTROL: Emery 17, go ahead.

19:49:46 [Sound similar to increased engine rpm]

19:49:47 [Recorded GPWS horn] *Whoop, Whoop, Pull up Whoop, Whoop, Pull up Whoop, Whoop, Pull up Whoop, Whoop, Pull up Whoop . . .*

19:49:47 FIRST OFFICER: Power.

19:49:51 . . . *Whoop, Pull up.* [GPWS recorded voice]

19:49:52 CAPTAIN: All right all right . . . all right.

19:49:54 FIRST OFFICER: Push.

19:49:54 FLIGHT ENGINEER: Okay, so . . . we're going back up. There you go.

19:49:58 CAPTAIN: Roll out.

19:50:01 [Sound of someone exhaling tensely]

19:50:04 CAPTAIN: Emery 17, extreme CG problem.

19:50:06 CONTROL: Emery 17, roger.

19:50:10 [Sound of someone exhaling tensely]

19:50:11 FLIGHT ENGINEER: Anything I can do, guys?

19:50:11 CAPTAIN: Roll out to the right.

19:50:12 FIRST OFFICER: Okay. Push. Push forward. Awww . . .

19:50:26 FLIGHT ENGINEER: You got the trim maxed?

19:50:28 FIRST OFFICER: Power.

19:50:28 FLIGHT ENGINEER: More?

19:50:29 FIRST OFFICER: Yeah.

19:50:29 *Whoop, Whoop, Pull up Whoop, Whoop, Pull up Whoop . . .*

19:50:32 FIRST OFFICER: We're gonna have to land fast.

19:50:32 . . . *Whoop, Pull up Whoop, Whoop, Pull up . . .*

19:50:36 CAPTAIN: Left turn.

19:50:36 FIRST OFFICER: Okay.

19:50:37 FIRST OFFICER: What I'm trying to do is make the aeroplane's position match the elevator. That's why I'm putting it in a bank.

19:50:45 CAPTAIN: All right.

19:50:45 FIRST OFFICER: Okay.

19:50:46 CAPTAIN: Left turn.

19:50:46 FIRST OFFICER: So we're gonna have to land it in like a turn.

19:50:47 CAPTAIN: Bring it around.

19:50:48 [Sound similar to stick shaker]

19:50:49 CAPTAIN: Bring it around.

19:50:49 FIRST OFFICER: God damn.

19:50:51 [Sound of grunt]

19:50:54 FIRST OFFICER: You got the airport?

19:50:56 CAPTAIN: Bring it around.

19:51:00 FIRST OFFICER: Power.

19:51:02 *Whoop, Whoop Pull up, Whoop.*

19:51:07 FIRST OFFICER: Power. Aww shit.

19:51:08 [Sound similar to impact]

19:51:09 END OF RECORDING

Flight 17 heavy crashed in a left-wing-low, nose-low attitude, into the Insurance Auto Auctions salvage yard, setting fire to 100–200 cars. Debris cut a swathe about 250 yards wide and 600 yards long. All three crew members died. According to the NTSB, 'The probable cause of the crash was a loss of pitch control resulting from the disconnection of the right elevator control tab. The disconnection was caused by the failure to properly secure and inspect the attachment bolt.'

LINNEUS, Maine, USA
19 July 2000

Airwave Flight 9807, a Grumman G-159 twin-prop
Gulfstream-I aircraft owned and operated by Airwave
Transport, departed Moncton Metropolitan Airport, New
Brunswick, Canada, at 11.48 p.m. for a courier flight to
Montreal. At 00.15:15 the aeroplane was assigned a
'block' altitude clearance between 14,000 and 15,000 feet.
At 00:25:21, the flight crew asked for and received clearance to climb to 16,000. At 00:30:12, the flight crew
declared an emergency and requested directions to the
nearest airport. The aircraft descended out of control and
crashed on the eastern side of the Meduxnekeag River.
It appeared that the crew experienced a failure of the
aeroplane's number one engine, prior to declaring an
emergency to air traffic control. The pilot-in-command
was the owner and president of Airwave Transport. He had
accumulated about 6000 hours of total flight experience, of
which 500 hours were as Gulfstream-I pilot-in-command.
The co-pilot had about 600 hours of total flight experience,
of which 300 were Gulfstream-I hours.

00:29:20 FIRST OFFICER: We got an engine failure,
number one.

00:29:27 CAPTAIN: Carry out the drill.

00:29:41 FIRST OFFICER: Feathered. RPM zero.

00:29:55 CAPTAIN: What the fuck is going on?

00:29:57 FIRST OFFICER: I don't know.

00:30:07 CAPTAIN: What is going on here?

00:30:09 FIRST OFFICER: I don't know . . . You're losing airspeed as well.

00:30:12 CAPTAIN: Okay. Declare an emergency.

00:30:25 FIRST OFFICER: Oh . . . keep it.

00:30:26 FIRST OFFICER: Keep it up. Keep it up.

00:29:16 [A sound similar to decrease in propeller rpm]

00:30:36 CAPTAIN: Oh no, uh oh.

00:30:42 CAPTAIN TO CONTROL: The co-pilot transmitted we've lost control.

00:.30:46 [A sound similar to varying change in propeller noise begins and continues to the end of recording]

00:30:51 CAPTAIN: Uh ohh.

00:30:54 FIRST OFFICER: Which way are we flying?

00:30:56 CAPTAIN: I have no –

00:30:56 FIRST OFFICER: I don't know I don't know.

00:31:09 CAPTAIN: I have no idea which way is up.

00:31:10 FIRST OFFICER: Oh. Ground . . . I don't know either.

00:31:13 CAPTAIN: . . . Upside down?

Both crewmen died.

The NTSB, which assisted in the investigation, concluded as a probable cause of the crash that 'the pilot-in-command [failed] to maintain minimum control airspeed, which resulted in a loss of control. Factors in this accident were clouds, and a loss of engine power for undetermined reasons, while in cruise flight above the aeroplane's single engine service ceiling.'

CHARLES DE GAULLE AIRPORT, Paris, France
25 July 2000

Air France Flight 4590, a Concorde registered F-BTSC, was to depart for New York-JFK Airport in the early afternoon of 25 July 2000. A crew request for a replacement of a part to the number two engine delayed departure by one hour. Finally, when all one hundred passengers had boarded, the plane taxied to runway 26R. Takeoff weight was calculated to be 186.9 tons, including 95 tons of fuel. At 14:42:17 the ground controllers cleared the crew for takeoff. At 14:42:54.6, the co-pilot called out 100 knots, then V one nine seconds later. As the Concorde sped down the runway, the right front tyre ran over a strip of metal lost by a Continental Airlines DC-10–30, registered N13067, which had departed Paris as Flight 055 to Newark five minutes earlier. The metal strip destroyed the Concorde's right tyre and in all probability a large piece of rubber from the disintegrating tyre was thrown against the underside of the left wing and probably ruptured a part of fuel tank five. A gout of flame spewed out under the left wing. Engines one and two lost thrust, severely in engine two and slightly in engine one.

As the Concorde continued to take off, the airport controller radioed the crew about the flames. The first officer acknowledged this transmission and the flight engineer announced the damaged to engine two. Nine seconds later the engine fire alarm rang and the flight engineer announced, 'Shut down engine two.' The captain called for the 'engine fire' procedure. A few seconds later, the crew pulled the fire handle for engine two fire handle. The co-pilot drew the captain's attention to the airspeed. The captain called for landing gear retraction. Ground control confirmed the presence of flames behind the aircraft. The engine fire alarm sounded again for around twelve seconds and then sounded for a third time and continued until the end of the flight. The first officer reported that the landing gear had not retracted and called out the airspeed. The GPWS sounded several times. The first officer informed air traffic control that they were trying to reach nearby Le Bourget Airport for an emergency landing. Then the number one engine lost power. The aircraft entered a left turn until the captain lost control and the Concorde crashed into the Hotel Hotellisimo and burst into flames, killing all aboard.

14:12:23 BEGINNING OF RECORDING
14:13:40 FIRST OFFICER: Fire detection.
14:13:42 FLIGHT ENGINEER: On, both.
14:13:43 FIRST OFFICER: Flight recorders with two hundred twenty.
14:13:44 FLIGHT ENGINEER: Two hundred twenty inserted, off.
14:13:46 FIRST OFFICER: Fire protection.
14:13:47 FLIGHT ENGINEER: Tested.
14:13:48 FIRST OFFICER: Pressurization.

14:13:49 FLIGHT ENGINEER: Six thousand . . .
[THE FLIGHT ENGINEER AND FIRST OFFICER
GO DOWN THE PRE-PUSHBACK CHECKS]
14:13:50 FIRST OFFICER: Engine control schedule.
14:13:55 CABIN ATTENDANT: Ladies and
gentlemen, good day. My name is [. . .], your chief flight
attendant. In the name of the captain . . .
14:14:09 FIRST OFFICER: Aircraft weight total fuel.
14:14:10 FLIGHT ENGINEER: One hundred
eighty-six ninety-five.
14:14:12 FIRST OFFICER: Totalizers.
14:14:14 FLIGHT ENGINEER: At zero.
14:14:15 FIRST OFFICER: Reference speeds.
14:14:23 CAPTAIN: So the reference speeds V one, one
hundred fifty, V R, one hundred ninety-eight, V two, two
hundred twenty-two one hundred forty-two one hundred
eighty . . .
14:14:43 FIRST OFFICER: Noise reduction.
14:14:45 CAPTAIN: The reduction is on seventy-three
seconds.
14:14:47 FLIGHT ENGINEER: Seventy-three.
14:14:48 CAPTAIN: Gilles, is that okay for you?
14:14:49 FLIGHT ENGINEER: Yes.
14:14:53 CAPTAIN: Then the thrust lever is at fourteen
and you will have N-2 of ninety-seven and a bit.
14:14:58 FLIGHT ENGINEER: Ninety-seven.
14:14:59 FIRST OFFICER: Parking brake.
14:15.00 CAPTAIN: It's on park . . .
14:15:09 FIRST OFFICER: The window.
14:15:11 CAPTAIN: It's closed locked on the left.
14:15:13 FIRST OFFICER: On the right.

14:16:03 FIRST OFFICER: Ground, hello, hello.

14:16:08 GROUND: Yes, I'm listening.

14.16.10 FIRST OFFICER: Where are we down there?

14:16:11 GROUND: Well, the mechanics have finished now. They're just getting off the last tool box now but the loading isn't quite finished yet.

14:17:22 FIRST OFFICER: As soon as they've finished with their mess . . .

14:17:23 FLIGHT ENGINEER: Yeah, and then after that we're going to start up.

14:17:36 GROUND: Okay, it's clear. You can go.

14:17:40 FIRST OFFICER: You have the controls.

14:25:45 CAPTAIN: Ladies and gentlemen, all is in order and we are starting up our engines.

14:25:51 GROUND: Air pressure established.

14:25:54 CAPTAIN: Perfect start-up three.

14:27:05 FIRST OFFICER: Concorde 4590, good evening. Echo twenty-six we will be ready to pull in two minutes.

14:27:14 CAPTAIN: Ready for [engine] two?

14:27:23 CONTROL: Air France 4590, good day. You are pushing back facing the west. Are you ready to push back?

14:30:02 GROUND: Okay, aircraft cleared ready to taxi.

14:30:05 CAPTAIN: So we have clearance. The brake is released. We can go back.

14:33:40 CAPTAIN: Concorde 4590, we will be ready to taxi in one minute. Goodbye.

14:33:48 CAPTAIN TO GROUND: Start-up complete

so you can stop listening in . . . Goodbye. Thanks for everything. Have a good day.

14:34:39 CONTROL: Air France 4590, good day. Taxi to holding point 26 right via Romeo 26 right.

14:35:32 CONTROL: Air France 4590, do you want Whisky ten or Romeo taxiway?

14:35:37 FIRST OFFICER: I need all the runway.

14:35:38 CONTROL: Okay. So you taxi for Romeo, Air France 4590.

14:35:58 FIRST OFFICER: Front wheel steering; apparently there isn't any.

14:36.00 CAPTAIN: Well . . . It's working. I'm managing to control the aircraft.

14:36:11 FIRST OFFICER: Nose droop at five.

14:36:27 CAPTAIN: I'll get up a bit of speed before trying the brakes.

14:36:29 FIRST OFFICER: Be careful. It's going fast.

14:39:04 CAPTAIN: So the takeoff is at maximum takeoff weight – 180 tons 100 which means four reheats with a minimum failure N-2 of ninety-eight. Between zero and one hundred knots, I stop for any aural warning . . .

14:39:25 CONTROL: Air France 4590, contact the Tower on 120.9.

14:39:29 FIRST OFFICER: 129, [Air France] 4590. Good afternoon.

14:39:35 CAPTAIN: Tyre flash and failure callout from you, right? Between 100 knots and V one, ignore the gong stop for an engine fire, a tyre flash, and the failure callout.

14:39:45 FIRST OFFICER: Yes.

14:39:46 CAPTAIN: After V one we continue on the SID we just talked about. We land back on runway 26 right. And the quick reference handbook is ready for an overweight landing.

14:39:56 CONTROL: Air France 4590, hello.

14:39:58 FIRST OFFICER: Hello. Threshold 26 right, [Air France] 4590.

14:40:01 CONTROL: [Air France] 4590, line up on runway 26 right.

14:40:04 FIRST OFFICER: We line up and hold on 26 right, [Air France] 4590.

14:40:07 CAPTAIN: Ready in the back?

14:40:10 FIRST OFFICER: Let's go.

14:40:11 CONTROL: Prepare for takeoff.

14:40:19 CAPTAIN: How much fuel have we used?

14:40:23 FLIGHT ENGINEER: We've got eight hundred kilos there.

14:40:24 CAPTAIN: Eight hundred kilos? Right?

14:40:28 FIRST OFFICER: No difference on the second segment.

14:41:04 FIRST OFFICER: Holding position, [Air France] 4590.

14:41:05 FLIGHT ENGINEER: The transponder. I'm putting it on.

14:42:17.0 TOWER: Air France 4590, runway 26 right, wind zero ninety-eight knots. Cleared for takeoff.

14:42:23.3 FIRST OFFICER: [Air France] 4590 takeoff 26 right.

14:42:24.8 CAPTAIN: Is everybody ready?

14:42:25.8 FIRST OFFICER: Yes.

14:42:26 FLIGHT ENGINEER: Yes.

14:42:26.6 CAPTAIN: To one hundred V one, one hundred-fifty.

14:42:30 [Noise of selector clicking of thrust levers]

14:42:31 [Change in background noise. Sound of increase in the airflow in the air conditioning and an increase in engine speed]

14:42:35 COCKPIT: Go Christian . . .

14:42:43.3 FLIGHT ENGINEER: We have four reheats.

14:42:54.6 FIRST OFFICER: One hundred knots.

14:42:57 FLIGHT ENGINEER: Four greens.

14:43:10.1 [Change in background noise. Unintelligible noise]

14:43:13 FIRST OFFICER: Watch out.

14:43:16.4 FLIGHT ENGINEER: Stop.

CONTROL: Concorde 4590, you have flames, you have flames behind you.

14:43:18.8 FIRST OFFICER: Roger.

14:43:20.4 FLIGHT ENGINEER: Failure eng, failure engine two . . .

14:43:22.8 [Bell fire alarm]

14:43:23.5 [Gong]

14:43:24.8 FLIGHT ENGINEER: Shut down engine two.

14:43:25.8 CAPTAIN: Engine fire procedure.

14:43:30 FIRST OFFICER: Watch the airspeed the airspeed the airspeed.

[Noise of selector]

[Gong]

[UNKNOWN SPEAKER] It's really burning and I'm not sure it's coming from the engines.

[Sound of fire handle being pulled]

14:43:32.6 FLIGHT ENGINEER: The gear.

[Toilet smoke detection alarm]

CONTROL: 4590, you have strong flames behind you.

14:43:34.5 CONTROL: Beginning of reception of a middle marker.

14:43:34.7 FIRST OFFICER: Yes, roger.

14:43:35.5 FLIGHT ENGINEER: The gear.

14:43:37 [Gong]

14:43.38.4

[Noise]

[Noise]

CONTROL: So do as you wish to have priority for a return to the field.

14:43:39 CAPTAIN: [Gear] retract.

14:43:41.2 FIRST OFFICER: Roger

14:43:42.3 [Bell fire alarm]

14:43:4 [Gong]

14:43:44.7 [Selector noises. Fire extinguisher fired with first shot]

14:43:45.6 FIRST OFFICER: I'm trying.

FLIGHT ENGINEER: I'm firing it.

14:43:46.3 CAPTAIN: [Are you] shutting down engine two there?

14:43:47.5 FLIGHT ENGINEER: I've shut it down.

14:43:49.9 FIRST OFFICER: The airspeed.

14:43:53 [Noise]

14:43.56.7 FIRST OFFICER: The gear isn't retracting.

14:43:58.6 [Bell fire alarm]

14:43:59.1 [*Whoop, Whoop, Pull up* . . .] [GPWS warning]

14:43:59.4 [Gong]

14:44:00 [*Whoop, Whoop, Pull up* . . .] [GPWS warning]

14:44:00.7 FIRST OFFICER: The airspeed.

14:44:02 [*Whoop, Whoop, Pull up . . .*] [GPWS warning]

14:44:03 FIRE SERVICE LEADER ON GROUND: De Gaulle tower from fire service leader.

14:44:05.2 CONTROL: Fire service leader, er . . . the Concorde . . . I don't know his intentions. Get into position near the southern double runway.

14:44:16.5 CAPTAIN: [Too late]

FIRST OFFICER: Le Bourget Le Bourget.

FIRE SERVICE LEADER: De Gaulle tower from fire service leader, authorization to enter 26 right.

14:44:19.8 CAPTAIN: [No, time no]

CONTROL: Fire service leader, correction. The Concorde is returning on runway 09 in the opposite direction.

14:44:22.8 FIRST OFFICER: Negative. We're trying for Le Bourget.

14:44:24.7 [Noise of selector]

[Noise of selector]

[Noise of selector]

[Noise of selector]

14:44:27 FIRST OFFICER: No.

[Gong]

[Gong and noise of selector]

FIRE SERVICE LEADER: De Gaulle tower from fire service leader, can you give me the situation of the Concorde now?

14:44:27.5 [Noise of selector and beginning of movement of objects in cockpit]

14:44:29 COCKPIT: [Noises of effort]

14:44:31.6 END OF RECORDING

ASPEN, Colorado, USA
29 March 2001

At 5.11 p.m. a Gulfstream-III, an executive jet registered as N303GA, owned by Airbourne Charter, Inc., and operated by Avjet Corporation of Burbank, California, departed Los Angeles International Airport (LAX) with two pilots, one flight attendant and fifteen charter passengers. While at LAX the captain and a charter department scheduler discussed the status of the passengers' arrivals and the weather currently being forecast for the flight's arrival time at Aspen. Also, the captain and another Avjet captain discussed the night-time landing restriction at Aspen Airport that required the aeroplane to land no later than thirty minutes after sunset in compliance with noise restrictions after 6.58 p.m. The late arrival of the charter passengers at LAX, including the charter customer, delayed the aeroplane's departure until 5.11, forty-one minutes later than originally scheduled. The flight's duration was to be 1 hour 35 minutes, so the estimated arrival time at Aspen was 6.46, a mere twelve minutes before the airport shut down for the night.

Interestingly, the client had chartered the Gulfstream private flight in order to be in Aspen that night to host a party. Presumably several of the aircraft's passengers were guests. Avjet called the business assistant of the chartering client at about 4.30 to tell him that the passengers were not yet at the airport and that the latest time the aeroplane could depart was 4.55. The assistant immediately began to track down the passengers and found that all but two (including the chartering client) were waiting in the airport parking lot. The charter department told the business assistant that the flight would instead have to change its flight plan and land at Rifle, an airport near Aspen, if the two passengers did not arrive shortly. According to the business assistant, the passengers in the parking lot waited in the aeroplane for the charter client and a couple of other passengers to arrive. One of the pilots spoke to one or more passengers, pointing out that the aeroplane might not be able to land at Aspen because of the night-time landing curfew. The charter client, upon learning of this conversation, instructed his business assistant (probably not in such polite language) to tell the pilot to keep his comments to himself. When the business assistant told his employer that the flight might have to divert, his employer became irate. The assistant was told to tell the company that the aeroplane was not going to be redirected. Why? The chartering client had flown into Aspen at night and was going to do it again that night, come hell or high water. The business assistant called Avjet to express his employer's displeasure. Later, during the flight, the captain stated, during an en route conversation at about 6.30, that it was important to land at Aspen because the customer had spent a substantial amount of money on dinner. The crew were under pressure, which would only increase, to land at Aspen, no matter what.

This client pressure served as a backdrop to an already pressured situation. Aspen Airport is at an elevation of

7815 feet and lies in a valley bordered by mountains. Approaches are difficult even in good weather and require considerable piloting skill, particularly when weather and winter storms limit visibility.

At 18:31:06 the captain stated, 'Well, there's the edge of night right here.'

At 18:31:24 the first officer asked the captain about the time of official sunset for Aspen.

The captain replied, 'Six twenty-eight', and then stated, 'So we get thirty minutes after sunset. So six fifty-eight . . . about . . . seven o'clock.'

At 18:37:04 the first officer called for the approach briefing. The captain then stated, 'We're . . . probably gonna make it a visual . . . If we don't get the airport over here we'll go ahead and shoot that approach', and 'We're not going to have a bunch of extra gas so we only get to shoot it once and then we're going to [the nearest airport to Aspen down valley] Rifle.'

The first officer acknowledged this information.

At 18:39:56 the flight crew began to receive automated information about the wind, visibility, sky condition and temperature information at Aspen, and the captain acknowledged this information.

At 18:44:22 the first officer made initial contact with Aspen approach control.

At 18:44:43 the flight crew heard, over the Aspen approach control frequency, the request of another private aircraft, a Canadair Challenger 600, for another approach to Aspen, after it had missed the first approach. The approach controller then cleared the Challenger to continue on a missed approach procedure.

At 18:45:00 the first officer stated, 'I hope he's doing practice approaches.'

About three seconds later, the captain asked the controller whether the pilot of the Challenger was practising or had actually missed the approach. The controller replied that the pilot had missed the approach. The controller also informed the captain that two other aeroplanes were on approach to Aspen.

At 18:45:32 the controller instructed the Gulfstream's flight crew to turn to a 360 degree heading to line up for the approach sequence.

While the aircraft was descending into Aspen, the flight crew discussed the location of a highway that runs parallel to the single runway at Aspen Airport.

At 18:45:45 the captain stated, 'Where's that . . . highway? Can we get down in there?'

About eleven seconds later he asked, 'Can you see?'

The first officer replied, 'I'm looking I'm looking . . . no.'

At 18:46:26, the captain said, 'I got it', and about two seconds later he asked the first officer, 'Can't really see up there, can ya?'

The first officer replied, over the next several seconds, 'Nope, not really', and 'I see a river but I don't see nothing else.'

At 18:47:19 the first officer said, 'I see . . . some towns over here and the highway's leading that way but I'm not sure.'

At 18:47:30 the approach controller made a blanket broadcast that the pilot of another private aircraft trying to land, a Cessna Citation, saw the airport at 10,400 feet and was making a straight-in approach.

The Gulfstream's first officer stated, 'Ah, that's good.'

(The Cessna Citation landed without incident at the airport about ten minutes before the Gulfstream crashed.)

At 18:47:41 the captain informed the controller, 'I can almost see up the canyon from here but I don't know the terrain well enough or I'd take the visual.'

About 18:47:51 the first officer said, 'Could do a contact [approach] but . . . I don't know', followed by 'Probably we could not . . .' The first officer also said, at 18:48:04, 'Remember that crazy guy in this Lear[jet] when we were . . . on the ground in Aspen last time and he [stated that he could] see the airport but he couldn't see it?'

About 18:48:51 the captain said, 'There's the highway right there.'

The first officer asked the captain if he wanted to be set up on the approach into Aspen, and the captain indicated that he was ready. He did not want to divert into Rifle unless he had to.

At 18:49:28 the captain asked the first officer whether he could see the highway. The first officer replied, 'No, it's clouds over here on this area. I don't see it.'

At 18:50:42 the captain said, 'But it's right there.' About six seconds later, he said, 'Oh, I mean we'll shoot it from here. I mean we're here but we only get to do it once.' (And if that approach failed, fuel and time would require them to divert to Rifle.)

At 18:50:54 the captain indicated to the flight attendant that if the attempt to execute the approach was not successful, they would have to go to Rifle because, 'It's too late in the evening then to come around.'

At 18:51:54, the approach controller instructed the flight crew to turn to a heading of 050 degrees.

At 18:53:09 the approach controller instructed the flight crew to turn to a heading of 140 degrees to intercept the final approach course and maintain an altitude of 16,000 feet. The controller then made a blanket broadcast that the last aeroplane, the Challenger 600, had missed the second approach, to which the Gulfstream's first officer remarked, 'That's . . . not . . . good.'

At 18:53:57 the flight attendant asked the cockpit crew whether a male passenger could sit on the jump seat in the cockpit. About eleven seconds later, the flight attendant instructed that passenger to make sure his seat belt was buckled.

At 18:55:05 the flight crew heard, over the approach control frequency, the pilot of the Challenger 600 transmit his intention to execute a missed approach. Afterwards the Gulfstream captain said, 'The weather's gone down. They're not making it in', and an unidentified male voice in the cockpit, presumably the jump-seat passenger, said, 'Oh really?'

At 18:56:06 the approach controller cleared the flight crew for an approach and advised the crew that the aeroplane was five miles from the initial approach fix, and instructed the crew to cross the VHF omnidirectional range beacon at or above an altitude of 14,000 feet. The first officer acknowledged this information.

At 18:56:23 the first officer said, 'After VOR you are cleared to twelve thousand seven hundred.'

At 18:56:42 the approach controller made a blanket broadcast that recorded weather information was current and that the visibility north of the airport was two miles. The approach controller then instructed the Gulfstream flight crew to contact Aspen local control, and the crew established contact with the local controller at 18:57:28. The local controller informed the flight crew members that they were following a Challenger aeroplane that was two miles from the runway, and reported the wind at 240 degrees at 5 knots, and cleared the aeroplane to land on runway 15.

At 18:57:55 the first officer acknowledged the clearance to land.

At 18:58:00 the local controller asked the Challenger pilot whether he had the airport in sight. He replied, 'Negative, going around.'

At 18:58:13 the jump-seat passenger in the cockpit asked, 'Are we clear?'

The captain replied, 'Not yet. The guy in front of us didn't make it either.'

At 18:58:27 the captain asked the first officer about the next step-down altitude, and he answered that it was 12,200 feet. The first officer also indicated that the next step-down fix was the beginning of the final approach segment.

At 18:59:11, the captain asked the first officer about the next step-down altitude, and the first officer answered 10,400 feet.

At 18:59:30 the first officer stated, 'Three greens.'

At 19:00:08, the jump-seat passenger in the cockpit said, 'Snow.'

At 19:00:22, when the aeroplane was at an altitude of 10,400 feet, the captain said, 'Okay . . . I'm breaking out [of clouds]', and asked the local controller, about five seconds later, whether the runway lights were turned all the way up. The controller said, 'Affirmative. They're on high.'

At 19:00:30, the first officer said, 'Okay, you can go . . . ten thousand two hundred [the minimum descent altitude].'

At 19:00:43 the captain asked the first officer whether he could see the runway. Two seconds later the first officer made an unintelligible reply.

At 19:00:46, with the aeroplane at an altitude of 10,000 feet, the captain asked the first officer whether he could see the highway, and he replied a second later, 'See highway.'

The local controller asked the flight crew members, at 19:00:49, whether they had the runway in sight. About two seconds later the first officer stated, 'Affirmative', and the captain stated, 'Yes, now yeah, we do.'

About one second afterwards, the first officer advised the controller that the runway was in sight.

At 19:01:13 the first officer stated, '. . . to the right is good.'

According to radar data, the airport was to the left of the aeroplane at this time.

At 19:01:28 the aeroplane's flight profile advisory (FPA) unit announced the 1000-foot callout, and the first officer stated, 'One thousand to go.'

At 19:01:31 and 19:01:34, the FPA called out 900- and 800-foot callouts respectively.

At 19:01:36 the captain asked, 'Where's it at?'

The FPA called out 700- and 600-foot callouts at 19:01:38 and 19:01:42 respectively. At 19:01:42 the first officer said, 'To the right', which the captain repeated about a second later. Radar data indicated that the airport was still to the left of the aeroplane at this point.

At 19:01:45, the aeroplane's GPWS and FPA units simultaneously announced the 500-foot callout. According to radar data, the aeroplane started a turn to the left at 19:01:47.

At 19:01:49 the GPWS announced a sink rate alert, and the FPA announced the 400-foot callout.

At 19:01:52 GPWS sink rate alert sounded and the FPA 400-foot callout sounded. The engines increased to maximum power at 19:01:53. The FPA 300-foot callout was sounded at 19:01:54, and the GPWS and FPA 200-foot callouts sounded about one and two seconds later respectively.

At 19:01:57 the GPWS's bank angle alert sounded when the aeroplane was banked about 40 degrees left wing down.

The aeroplane crashed into terrain while in a steep left bank about 2400 feet short of the runway 15 threshold, 300 feet to the right (west) of the runway centreline and 100 feet above the runway threshold elevation.

The CVR stopped recording at 19:01:58.

The NTSB determined that the probable cause of this accident was the flight crew's operation of the aeroplane below the minimum descent altitude without an appropriate visual reference for the runway. Contributing to the cause of the accident was the inability of the flight crew to adequately see the mountainous terrain because of the darkness and the weather conditions, pressure on the captain to land from the charter customer, the aeroplane's delayed departure and the airport's night-time landing restrictions.

All three crew members and fifteen passengers died.

NEAR BURDAKOVA, Russia
4 July 2001

At 7.47 p.m., Vladivostokavia flight 352, a Tupolev 154M, registered as RA-85845, took off from Ekaterinburg-Koltsovo, Russia, on a flight to Vladivostok, Russia, with an intermediate stop at Irkutsk, Siberia, Russia, carrying 145 crew and passengers. The flight climbed to the cruising altitude of 33,136 feet. Some six hours after takeoff, at 1.50 a.m., the aircraft descended for an approach into Irkutsk. At 2.05 a.m. the crew reported the runway in sight while flying at 6890 feet and at 335 mph. The maximum speed at which to safely lower the landing gear was 248 mph. At 02:06:56 the aeroplane levelled off at 2952 feet and at about 335 mph. The first officer ordered the landing gear down and the speed further reduced to 245 mph with engines at idle. When the gear was down and locked the aeroplane entered a left-hand 20–23 degree turn. The airspeed decreased to 226 mph. Power was added slowly to maintain an altitude of 2788 feet at 220–224 mph. At 02:07:46, while still in the left bank, the autopilot attempted to maintain altitude by increasing the aircraft's angle of attack. An aural warning indicated a high angle of attack which the first officer attempted to correct by pulling back on the yoke. Due to these human inputs, the left bank increased and the nose pitched down; the speed increased to 248 mph. The aeroplane entered a layer of clouds. The captain turned the steering yoke left and right and the aircraft banked sharply left (-45 degrees). The vertical descent rate increased. Either the captain or first officer pulled the control column back. The aeroplane pitched up rapidly and stalled before entering a flat spin from which the crew could not regain control. After spinning for twenty-two seconds, the aircraft slammed down on its belly, broke up and burned.

01:58:10 CONTROL: 845, Razdolye [waypoint] confirmed 5700 [metres] contact approach 125.20.
01:58:21 CAPTAIN: Okay.
01:58:22 CAPTAIN TO CONTROL: Irkutsk approach 85845, g'evening Razdolye at 5700, information X-ray.
01:58:32 CONTROL: 85545, Irkutsk approach, g'evening, final 282 degrees, distance 80 [kilometres], descend and maintain 2700 [metres] to downwind, left pattern.
01:58:43 CAPTAIN TO CONTROL: 845, to downwind descending 2700 metres.
01:59:25 CAPTAIN: Sounds good.
01:59:35 CAPTAIN: . . . it will take us 550 metres.
01:59:46 CAPTAIN: To downwind.
02:01:04 CAPTAIN: It seems we have to put it up.
02:01:43 FLIGHT ENGINEER: Switched on.
02:01:55 CONTROL: 845, distance 40 [kilometres], descend and maintain 2100 [metres].
02:01:58 FIRST OFFICER: 2100 [metres].
02:01:59 CAPTAIN TO CONTROL: 845, descending 2100 [metres].

02:02:59 CAPTAIN: At home while descending on straight-in approach, we're passing DAGES on that altitude. Here we've got more 550 metres. We'll drop it on. [DAGES is name of intersection]

02:04:09 CONTROL: 845, latest Yankee [ATIS], confirm you've got Yankee.

02:04:13 CAPTAIN TO CONTROL: Roger, listening now.

02:04:52 CONTROL: 845, what type of approach?

02:04:55 CAPTAIN TO CONTROL: NDB approach, information Yankee, altimeter 7–10.

02:05:02 CONTROL: 845, you're cleared for NDB approach, descend 900 [metres] for base altimeter 7–10.

2:05:08 CAPTAIN TO CONTROL: 845, descending 900 [metres] for base turn, altimeter 7–10.

02:05:14 CONTROL: 845, you're downwind now, 11 kilometres off final.

02:05:18 CAPTAIN TO CONTROL: Roger, we have [airport] in sight, that's copied.

02:05:29 CAPTAIN: Yep, it is working.

02:05:32 FIRST OFFICER: I'm expecting 20 kilometres.

02:05:34 CAPTAIN: Okay.

02:05:36 FIRST OFFICER: Altimeter now 7–10 of mine.

02:05:43 CAPTAIN: . . . and slow down, decelerate speed.

02:05:44 CAPTAIN: Decelerate, decelerate.

02:05:47 CAPTAIN: Let's set altimeter 7–10.

02:05:48 FIRST OFFICER: 7–10.

02:05:55 CAPTAIN TO CONTROL: 845, at 1800

[metres], altimeter set 7–10, will report through 1250 [metres], descending 900 [metres].

02:06:01 CONTROL: Roger.

02:06:05 CAPTAIN: Decelerate, reduce speed.

02:06:05 FIRST OFFICER: Reducing, reducing.

02:06:08 CAPTAIN: We've got 100 metres to assigned [altitude].

02:06:09 FIRST OFFICER: That's affirmative, 100 metres.

02:06:13 CAPTAIN: You've set 8 [8 metres per second descend rate]. That's why you've got such speed.

02:06:17 CAPTAIN: Here we go, Yura. Now in clouds.

02:06:20 CAPTAIN: Look there, what temperature is.

02:06:21 FIRST OFFICER: They gave 290 degrees at 5, 14 degrees, 710 CAVOK.

02:06:25 CAPTAIN: All right.

02:06:26 CAPTAIN: Okay, now turn it on and . . .

02:06:28 FIRST OFFICER: Is it engaged? Is light on?

02:06:30 FIRST OFFICER: . . . is on [working]?

02:06:32 CAPTAIN: How much remaining?

02:06:34 FIRST OFFICER: 8 kilometres.

02:06:35 CAPTAIN: Copied.

02:06:56 CAPTAIN: Okay, take it there, I've switched hold altitude.

02:06:58 FIRST OFFICER: Okay.

02:07:02 FIRST OFFICER: 400 kilometres [km/hour].

02:07:03 CAPTAIN: Rudder good, gears down.

02:07:06 FIRST OFFICER: Putting [gear] down.

02:07:08 FLIGHT ENGINEER: They're going [gear's going down].

02:07:10 FIRST OFFICER: Gears . . .

02:07:12 CONTROL: 845, turn to base, descent 850 metres for final.

02:07:17 CAPTAIN TO CONTROL: 845, descending 850 [metres] for final.

02:07:21 FLIGHT ENGINEER: Gear's down.

02:07:23 CAPTAIN: Okay, speed is decreasing too much.

02:07:25 CAPTAIN: Descend.

02:07:27 CAPTAIN: 850 [metres], power 70.

02:07:29 FLIGHT ENGINEER: 70.

02:07:30 CAPTAIN: 350 [km/hour], speed in range.

02:07:32 CAPTAIN: Power 7–5.

02:07:33 FLIGHT ENGINEER: 7–5.

02:07:37 FIRST OFFICER: Approaching 850 [N1metres].

02:07:39 CAPTAIN: Power 80.

02:07:40 FLIGHT ENGINEER: 80.

02:07:42 CAPTAIN: 8–2.

02:07:44 FIRST OFFICER: 850 [metres].

02:07:45 CAPTAIN: 850 [metres].

02:07:45 Audio tone of AOA [angle of attack] alert and maximum acceleration reached [duration 1.5 seconds]

02:07:47 Audio tone of autopilot disconnection warning [duration 2.3 seconds]

02:07:49 CAPTAIN: Fuck. What are you doing?

02:07:51 CAPTAIN: Speed.

02:07:53 CAPTAIN: . . . Fuck, push it [throttles] up.

02:07:53 FIRST OFFICER: Stop. Stop. Where. Where.

02:07:55 CAPTAIN: Stop. Stop. Stop.

02:07:55 FIRST OFFICER: This way, this way, this way.

02:07:57 CAPTAIN: We're recovering.

02:07:58 FIRST OFFICER: Easy, make it easy, easy.

02:07:59 FIRST OFFICER: Let's to the right.

02:08:01 Tone of radio-altitude alert [duration four seconds] and audio alert tone of AOA [remaining to the end of recording].

02:08:02 COCKPIT: Power. Add thrust.

02:08:05 COCKPIT: Power!

02:08:06 FLIGHT ENGINEER: . . . got it.

02:08:08 COCKPIT: Add thrust!

02:08:09 FIRST OFFICER: Takeoff power. Oh, my God.

02:08:10 FLIGHT ENGINEER: Takeoff power set.

02:08:11 COCKPIT: That's all guys. Fuck.

02:08:16 Audio tone of radio altimeter alert [remaining to the end of recording].

All 145 occupants of the aircraft died.

BELLE HARBOR, New York, USA

12 November 2001

On 12 November 2001, at around 9.16 a.m., American Airlines Flight 587, an Airbus MA300–605R, N14053, crashed into a residential area shortly after takeoff from John F. Kennedy International Airport. Flight 587 was a regularly scheduled passenger flight to Santo Domingo, Dominican Republic, with two flight crew members, seven flight attendants and 251 passengers aboard the aircraft.

The aeroplane had arrived at JFK at about 10.31 p.m. the night before from San Jose, Costa Rica, with an intermediate stop in Miami. The pilots of the leg from Miami to New York indicated that the flight was smooth and uneventful. Flight 587 was to be the first leg of a one-day round-trip sequence for the flight crew. The captain checked in for the flight at about 6.14 a.m. and the first officer at 6.30. The gate agent working the flight arrived at the departure gate at about 6.45. The flight attendants were already aboard at that time and the captain and the first officer arrived at the gate at about 7.00 a.m. Of the 251 passengers, five were lap children (i.e. a child without a seat or seat belt of its own) under two years of age.

At about 7.10 a.m. the aeroplane fuelling process began. The aeroplane fueller indicated that he saw one of the pilots perform an exterior inspection of the aircraft. He finished the fuelling process at about 7.45 and saw nothing unusual regarding the aeroplane.

Statements were provided to the Port Authority of New York.

Between 7.30 and 8.00 a.m. the captain reported that the number two pitch trim and yaw damper system would not engage. Two avionics technicians investigated the problem, performing an auto flight system (AFS) check that indicated a fault with the number two flight augmentation computer. The circuit breaker was then reset, another AFS check was performed, and no fault was detected. An auto/land system check was performed, and that test also did not detect a fault. The avionics technicians were in the cockpit for five to seven minutes.

The cockpit voice recorder (CVR) recording began at about 8.45 a.m.

At about 9.02 a.m. the captain told the first officer, 'Your leg, you check the rudders.' (The first officer was the PIC, pilot-in-control, and the captain was the non-flying pilot.) At about 9.06 the ground controller provided the pilots of Japan Air Lines Flight 47, a Boeing 747–400, with taxi instructions to runway 31L. At about 9.08 the ground controller instructed the Japan Air Lines pilots to contact the local (tower) controller. A minute later the ground controller instructed American Airlines Flight 587's pilots to follow the Japan Air Lines aeroplane and to contact the local controller. The American first officer acknowledged this instruction. At about 9.11 the local controller cleared the Japan Air Lines aeroplane for takeoff and seconds later cautioned the American Flight 587 pilots about wake turbulence (turbulence that forms behind an aircraft as it passes through the air; behind a large, heavy aircraft the turbulence can be powerful enough to roll or even break up a smaller aircraft) and instructed the American pilots to taxi into position and hold for runway 31L. At about 9.13, after the Japan Air Lines flight had taken off, American 587's captain said to the first officer, 'You have the aeroplane.' At 9.13 the local controller cleared American Flight 587 for takeoff, and the captain acknowledged the clearance. At about 9.13 the first officer asked the captain, 'You happy with that [separation] distance?'

The captain replied, 'We'll be all right once we get rollin'. He's supposed to be five miles by the time we're airborne, that's the idea.' At 09:13:46 the first officer again asked, 'So you're happy?' He was clearly concerned about the wake turbulence from the Japan Air Lines flight. American Flight 587 started its takeoff roll and lifted off about 1 minute 40 seconds after the Japan Air Lines aeroplane.

At 09:14:43 the local controller instructed American Flight 587 to turn left . . . and contact the New York departure controller. About five seconds later, the captain acknowledged this instruction. Radar data indicated that the aeroplane climbed to 500 feet and then entered a climbing left turn to a heading of 220 degrees. At 09:15:00 the DS informed departure control that the aeroplane was at 1300 feet and climbing to 5000 feet. At 09:15:05 the departure controller instructed American Flight 587 to climb to and maintain 13,000 feet, and the captain acknowledged this instruction about five seconds later. At 09:15:29 the CVR recorded the captain's saying, 'Clean machine', indicating that the gear, flaps and slats had all been retracted. At 09:15:35, American Flight 587 was climbing through 1700 feet with its wings approximately level. One second later, the departure controller instructed the pilot in control to turn left and proceed direct to a navigation intersection located about thirty miles south-east of the airport. At 09:15:41 the captain acknowledged the instruction. The controller heard nothing further from Flight 587.

The flight data recorder data indicated what happened. At 09:15:36, the aeroplane experienced wake turbulence. Between 09:15:36 and 09:15:41, the pilot in control moved the control column, control wheel and rudder pedals, presumably to correct for the wake turbulence. At 09:15:44.7 the captain stated, 'Little wake turbulence, huh?' to which the first officer replied, 'Yeah.' At 09:15:48.2 the first officer indicated that he wanted the airspeed set to 250 knots, which was the maximum speed for flight below 10,000 feet. The aeroplane was at an altitude of about 2300 feet. At 09:15:51, 09:15:52.3.0 and 09:15:52.9 respectively, the CVR recorded the sound of a thump, a click and two thumps. At 09:15:54.2 the first officer stated, in a strained voice, 'Max power.' At that point the aeroplane was travelling at 240 knots. The captain asked, 'You all right?' The first officer replied, 'Yeah, I'm fine.' One second later, the captain stated, 'Hang onto it. Hang onto it.'

We pick up the CVR of American Flight 587 while it was still at the gate.

08.45:50 [Sound similar to paper rustling]

08:46:04 [Unintelligible comment]

08:46:05 FIRST OFFICER: Now what?

08:46:44 CAPTAIN: . . . seagulls getting in . . . Why are they flying around that construction site?

08:46:50 CAPTAIN: Oh, coffee truck.

08:46:51 FIRST OFFICER: [Sound of chuckle] . . . flying around 'cause it looks like a dump. That's why. They don't know any better. Good question. This thing's going triple the speed it was. Did you see that thing the union passed just before they approved the whole thing? They changed something with the list or something.

08:47:24 CAPTAIN: St Louis thing.

08:47:26 FIRST OFFICER: What did they do?

08:47:28 CAPTAIN: Ah, you know . . . we were talking about this last time, I guess you know . . . was [always saying] how I, I was getting the, the information, and, um, I guess there were some, uh, holes in the previous document, where the pilots of TWA guys in St Louis given a certain scenario could cross through the fence and come out of the AA system as captain's.

08:48:00 FIRST OFFICER: Eeewh.

08:48:03 CAPTAIN: Beyond that I, I don't, I can't really explain it . . . uh, but my understanding is, they, they plugged those holes in the fence.

08:48:13 FIRST OFFICER: 'kay.

08:48:20 CAPTAIN: Sit there and read this, this legalese stuff . . . I mean, you really have to sit down and study this to get it.

08:48:25 FIRST OFFICER: Really hard. Oh, yeah. And it's also a moot point seeing how it's passed now.

187

08:48:35 CAPTAIN: Yeah.

08:48:39 FIRST OFFICER: Which is good, which means . . . was causing trouble and He . . . plugged the hole.

08:48:44 CAPTAIN: You know he's a force now. He ought to run for, ah, union job. Boy, he's a New Yorker with an attitude.

08:49:02 FIRST OFFICER: We need a new . . .

08:49:04 CAPTAIN: Yeah. Well, [—is] irreplaceable. What was he saying to you this morning, anything?

08:49:11 FIRST OFFICER: Nothing.

08:49:12 CAPTAIN: 'cause he's still on the inside, isn't he?

08:49:14 FIRST OFFICER: Oh yeah. He's a . . . he's on the inside, very much so . . . he still goes down, you know . . . doing, you know . . . whatever . . . He's very much on the inside. Trust me, when friggin' has a problem he goes . . . this guy did this or whatever . . . got a friggin' wealth of knowledge sitting there. It's like . . . except that there's a wealth of knowledge on how to give things back, which we shouldn't have been tapping.

08:49:54 CAPTAIN: What?

08:49:57 FIRST OFFICER: . . . was calling up . . . You know we were paying . . . ? This [damned] union didn't even tell us way back then.

08:50:02 CAPTAIN: Is that right?

08:50:03 FIRST OFFICER: Oh yeah. [Unnamed person] was hired as a consultant. And taken off flight [status] . . . No one told us about that.

08:50:15 CAPTAIN: I didn't know that. So they, they really didn't tell us.

08:50:18 FIRST OFFICER: Oh no.

08:50:19 CAPTAIN: They still haven't told us. We just found out.

08:50:21 [Sound of hiccup and cough]

FIRST OFFICER: Excuse me, I was just reading uh, ah . . . I think I found it out in the accounting thing. I was scouring through, um, APA's accounting [and] all the people that got flight-time pay and how much it was. You know and that's where I found it, that [he] got removed from, trip pay. Paid by APA, which could only be one, I mean, what else is he doing?

08:50:50 CAPTAIN: Yeah.

08:51:03 [Sound similar to yawn]

08:51:11 [Sound of singing]

08:51:24 FIRST OFFICER: I think across from it, it said consulting. [Sound of chuckle]

08:51:27 CAPTAIN: Said what?

08:51:28 FIRST OFFICER: Said removed from trip. It said like 'reason' or whatever. It said 'consulting' or something. Well, we're getting paid, that's nice. Thank you very much. Can't beat that.

08:52:44 CAPTAIN: The door is closed.

08:52:53 INTERCOM: Ground to cockpit?

08:52:54 CAPTAIN: Hello.

08:52:55 GROUND: Hello, cockpit. We just locked up and we're all secure below, [and] standing by.

08:52:58 FIRST OFFICER: American 587, ready to do push-back. Okay, brakes released. Stand by for the clearance.

08:53:03 CONTROL: 587, stand by. You're gonna be number two to push. I'll give you a call.

FIRST OFFICER: Uh, probe heat.

08:53:11 CAPTAIN: On.

08:53:11 FIRST OFFICER: ECAM doors display, slides? [Electronic centralized aircraft monitor is a system that monitors aircraft functions and relays them to the pilots. It also produces messages detailing failures and in certain cases lists procedures to undertake to correct the problem]

08:53:12 CAPTAIN: Green and armed.

08:53:13 FIRST OFFICER: Beacon, nav lights?

08:53:14 CAPTAIN: On, on.

08:53:15 FIRST OFFICER: Cabin ready?

08:53:15 CAPTAIN: Received.

08:53:18 FIRST OFFICER: Checklist is done. We're not cleared.

08:53:21 CAPTAIN: Okay.

08:53:24 FIRST OFFICER: Number two.

08:53:26 CAPTAIN: Three four and all right five thousand pounds heavy? Hm. I guess there's traffic out there somewhere . . .

08:55:57 PUBLIC ADDRESS SYSTEM: Well, ladies and gentlemen, we're all buttoned up ready to go. We're just waiting for an aeroplane behind us . . . uh, to move on out of our way, and then we will be pushing back.

08:56:07 [Sound of hi-lo chime]

08:56:08 [Sound of male flight attendant beginning passenger announcement in Spanish]

08:56:08 RAMP: American 587 . . .

08:56:09 FIRST OFFICER: Yes.

08:56:11 RAMP: 587, do you still have a ground crew there?

190

08:56:12 FIRST OFFICER: I believe we have ground crew. Yes, we have ground crews still.

08:56:15 RAMP: American 587 [Have your ground crew to reference the United seven thirty seven departing the alley]. You're cleared to push.

08:56:16 CAPTAIN: Huh? What?

08:56:20 FIRST OFFICER: Reference the seven three [Boeing 737]. We're cleared to push, American 587 heavy.

08:56:30 CAPTAIN: Evidently there's a seven thirty-seven back there. Uh, once you see him, we're cleared to push.

08:56:31 RAMP: Okay cockpit . . . um, they just disconnected and should be another couple of minutes.

08:56:37 CAPTAIN: Okay, whatever you like.

08:56:49 FIRST OFFICER: I can't believe how much money GE makes just renting little shacks to construction people. You know if GE's in, uh, it's huge money. I mean they don't, they don't . . .

0856:56 CAPTAIN: They're, they're, you know, if you looked into it, you could probably find GE in virtually everything.

08:57:02 FIRST OFFICER: GE is actually like one of the largest banks in the world. It's like the largest bank in the world.

08:57:05 CAPTAIN: Yeah.

08:57:07 FIRST OFFICER: They have more flexibility because they don't have the restrictions of a bank. Banks have certain restrictions as to what they can and can't do as far as, their hands tied. It's unbelievable. They own more aircraft than American Airlines.

08:57:25 CAPTAIN: GE does?

08:57:32 FIRST OFFICER: I'm not gonna say size wise but a lot of those corporate jets and all that other stuff . . . more aircraft.

08:57:37 CAPTAIN: Just numbers, yeah.

08:58:00 RAMP: Hey, cockpit, that aircraft is clear, we'll start our pushback.

08:58:04 CAPTAIN: Brakes released, cleared to push . . .

08:59:27 GROUND: Ground to cockpit, you're cleared to start.

08:59:29 CAPTAIN: Cleared to start [engines].

08:59:34 CAPTAIN: Starting two. Valves open. Brakes are parked, cleared to disconnect, see you out front.

08:59:58 GROUND: Ground to cockpit, disconnecting, see ya out front.

09:00:00 CAPTAIN: So long. The valve is closed. Starting [engine] one. Valve's open. See ya salute, four guys going away.

09:01:15 RAMP: 587, cleared to contact ground, have a good flight.

09:01:20 FIRST OFFICER: Good day. Morning ground, it's American 587 heavy with the information. Delta coming out of, uh, Tango Alpha.

09:01:30 FIRST OFFICER: Clear right.

09:01:31 CAPTAIN: Clear left. Fifteen when you're ready.

09:01:33 GROUND: American 587 heavy, Kennedy ground, runway 31 left for departure. Taxi left on Bravo. Hold short of Juliet.

09:01:39 FIRST OFFICER: Left Bravo, short of Juliet, American 587 heavy.

09:02:05 CAPTAIN: [It's] your leg. You check the rudders.

09:02:23 FIRST OFFICER: Rudders check.

09:02:26 CAPTAIN: Okay.

09:02:40 FIRST OFFICER: Taxi checklist is complete. Takeoff checklist, anti-ice?

09:03:01 CAPTAIN: Off.

09:03:02 FIRST OFFICER: Autobrakes.

09:03:03 CAPTAIN: Max.

09:03:38 FIRST OFFICER: Okay, the box is updated. We have, stand alone sheet, for . . . runway 31 left. Flaps fifteen, bleeds on. Assumed temperature is supposed to be forty-two. We have forty-two set. Weight was three forty-nine point three. And your stand alone sheet's for three forty-nine three. Weights check, temperature check. And I'm gonna double-check the winds here. Forty-two degrees is supposed to give us one oh one point one. We got one oh one point one. Numbers are one fifty, fifty four and fifty six . . . three-thirty at eleven. Winds checked. Takeoff data and TRP for thirty-one left.

09:04:28 CAPTAIN: Two hundred forty-six people, crew of nine, two hundred fifty-five SOBs. Takeoff data, set and cross-checked, flex power. Three one left, Kennedy.

09:04:36 FIRST OFFICER: Set and cross-checked. Takeoff data and TRP. AFS panel and radios?

09:04:41 CAPTAIN: Ah, fifty-six is preset, two forty-two, everything else remains the same. Set checked.

09:04:45 FIRST OFFICER: Gotcha. Flaps fifteen, stab trim is nose up point seven . . . stab trim?

09:04:53 CAPTAIN: Uh, point seven nose up set.

09:04:56 FIRST OFFICER: Slats and flaps?

09:04:57 CAPTAIN: Calls for fifteen, fifteen, set at fifteen, fifteen, fifteen. Fifteen, uploaded.

09:05:01 FIRST OFFICER: Verified fifteen, fifteen. Takeoff config?

09:05:09 CAPTAIN: Norm . . . takeoff.

09:05:11 FIRST OFFICER: Takeoff briefing? All right, you start out. If something happens prior to V one, call what you see, I'll decide whether or not to abort. Uh, unless it [is] an engine failure or an inability to fly [and] we'll plan on continuing the takeoff. Uh, it's a hundred feet and then, uh, a left turn to what?

09:05:40 CAPTAIN: Uh, runway heading to three hundred feet, that's a heading to two one zero to a thousand feet and then it's the, uh, engine, uh, clean-up or the aeroplane clean-up stuff . . . after the immediate action items you have the aeroplane. The radio we'll plan on left-hand traffic to come back to, uh, either one of the three ones . . . highest min safe altitude on runway heading is twenty-eight hundred foot, and once you make a left turn over water it's eighteen hundred feet. The terrain is flat with towers. Otherwise you plan on about a heading two two zero, five thousand feet. Questions?

09:06:15 FIRST OFFICER: No.

09:06:16 CAPTAIN: Briefing complete . . .

09:06:53 GROUND: Japan Air 47, continue via Bravo, turn right at Juliet, cross runway 4 left.

09:07:00 JAPAN AIRLINES: JAL Japan Air 47, on Bravo, Juliet cross runway 22, uh, 4 left.

09:07:07 GROUND: Japan Air 47, that's correct, thanks.

09:07:55 FIRST OFFICER: I was flying in here about three nights ago, comin' in, 'bout ten o'clock, doing that twelve, twelve, thirty departure turn, Dingo turn. So I don't know comin' in here, not ten o'clock somewhere there, nine o'clock, somewhere, and uh . . .

09:08:01 GROUND: Japan Air 47 heavy, monitor the tower one one niner point one. So long.

09:08:05 JAPAN AIRLINE: One one nine one, Japan Air 47, so long.

09:08:12 FIRST OFFICER: EgyptAir was told to do. They were rocketing off towards the city and they were told to . . .

09:08:17 CAPTAIN: EgyptAir was told to do what?

09:08:19 FIRST OFFICER: Turn thirty degrees, somewhere you know, like, you know, [they were] thirty degrees off their course. They were told to pick up a heading, and he said 'roger' and he didn't turn. And the controller said, 'Pick up this heading', and he didn't turn. [The controller] says, 'You need to turn immediately now EgyptAir, and I wanna know why you're not turning.' Finally he turned. [The ground controller] says, 'EgyptAir, we need to discuss on the ground why it took you fifteen miles to make a heading change when I asked you and you responded.' They were really . . .

09:08:47 CAPTAIN: Pissed.

09:08:48 FIRST OFFICER: . . . Pissed, and, uh, I think they were like ready to . . .

09:08:51 CAPTAIN: Launch the fleet?

09:08:52 FIRST OFFICER: Oh yeah, 'cause he was heading towards [New York City].

09:08:54 CAPTAIN: Oh really.

09:08:56 FIRST OFFICER: Oh yeah, that's why he's ticked, that's why he's so ticked.

09:08:58 GROUND: American 587 heavy, follow the Japan Air heavy Boeing 747 ahead. Monitor the tower one one niner point one. So long.

09:09:03 FIRST OFFICER TO GROUND: Follow Japan Air over to tower nineteen one, American 587 heavy. [He was] really ticked.

09:09:13 CAPTAIN: Follow JAL.

09:09:17 FIRST OFFICER: These guys just, uh, merged with Japan and another . . . what's the other Japan company?

09:09:20 CAPTAIN: All Nippon?

09:09:21 FIRST OFFICER: No, not Nippon. There's another one, I think. They merged this morning.

09:09:27 CAPTAIN: Really.

09:09:27 FIRST OFFICER: [Sound of yawn] Yeah. The big news, Japan and what other Japanese airline is . . . I don't think it was All Nippon. It was uh . . .

09:10:15 CAPTAIN: Crossing 22 right. Clear on the left.

09:10:20 FIRST OFFICER: On the right.

09:10:27 TOWER: Japan Air Lines 47 heavy, Kennedy tower, runway 31 left, taxi into position and hold.

09:10:30 JAL47: Runway 31 left, taxi into position and hold.

09:10:34 TOWER: PD fourteen, uh, caution wake turbulence. There'll be, uh, several heavy jets departures over Canarsie momentarily.

09:10:41 PD14: Roger that PD fourteen. We'll be looking.

09:10:44 [Sound of clunk]

09:10:51 PUBLIC ADDRESS SYSTEM: Well, ladies and gentlemen, at long last, we are number two for takeoff. Uh, towards the north-west today. Immediately after takeoff we'll be in a left-hand turn heading for the shoreline and, uh, getting ourselves pointed southbound. 'Bout another two or three minutes it'll be our turn to go. Flight attendants, prepare for takeoff, please.

09:11:08 TOWER: Japan Air Lines 47 heavy, wind three zero zero at one zero runway 31 left, cleared for takeoff.

09:11:12 JAPAN AIR LINES: Runway 31 left, cleared for takeoff, Japan Air 47 heavy.

09:11:27 CAPTAIN: Yeah, I guess that controller was bent outta shape, huh?

09:11:29 FIRST OFFICER: Ticked.

09:11:33 CAPTAIN: Can't hardly blame him.

09:11:35 FIRST OFFICER: Ah, I'm sure.

09:11:36 TOWER: American 587 heavy Kennedy tower, caution wake turbulence runway 31 left. Taxi into position and hold.

09:11:41 FIRST OFFICER: Position and hold three one left, American five eighty-seven heavy.

09:11:44 CAPTAIN: Position and hold. I see traffic out there. Hopefully he's going to the right side.

09:11:55 FIRST OFFICER: Takeoff briefing we got, PA?

09:11:57 [Sound of single chime]

09:11:58 CAPTAIN: Complete.

09:12:00 FIRST OFFICER: Ignition.

09:12:05 CAPTAIN: Reach? Ignition's off.

09:12:07 FIRST OFFICER: Bleeds?

09:12:09 CAPTAIN: Bleeds are on. He say Reach?

09:12:13 FIRST OFFICER: Reach four oh one or something like that, yeah.

09:12:16 CAPTAIN: Air Force is coming to Kennedy.

09:12:19 FIRST OFFICER: Is that a [call sign] Reach? That's the Air Force?

09:12:20 CAPTAIN: Yeah, that's the . . .

09:12:20 FIRST OFFICER: Transponder?

09:12:22 CAPTAIN: It's a tanker. The call sign. That looks like a seven four out there though. All right, position and hold on the, uh, left side. Final appears clear, transponder is TARA . . .

09:12:38 FIRST OFFICER: Brake . . . thank you. Brake fans?

09:12:39 CAPTAIN: Fans are off.

09:12:40 FIRST OFFICER: Lights to go. I'm gonna make . . . left turn two twenty. Go out the Bridge [at] five thousand feet's the top. If we have a problem, I'll clean it up at six [thousand feet] . . . ten . . . left traffic for this runway . . .

09:13:05 TOWER: Japan Air 47 heavy, fly the Bridge Climb, contact New York departure, good morning.

09:13:10 JAPAN AIR: Bridge Climb, switch to departure, Japan Air 47, good morning.

09:13:21 CAPTAIN: You have the aeroplane.

09:13:21 FIRST OFFICER: I got the brakes.

09:13:22 CAPTAIN: I have the radios.

09:13:27.6 TOWER: American 587 heavy, wind three zero zero at niner, runway 31 left, cleared for takeoff.

09:13:31.7 CAPTAIN: Cleared for takeoff, American, ah, 587 heavy.

09:13:35.3 FIRST OFFICER: You happy with that distance [for wake turbulence]?

09:13:38.5 CAPTAIN: Ah, he's . . . we'll be all right once we get rollin'. He's supposed to be five miles by the time we're airborne, that's the idea.

09:13:45.5 FIRST OFFICER: So you're happy. Lights?

09:13:47.1 CAPTAIN: Yeah, lights are on.

09:13:47.8 FIRST OFFICER: Takeoff check's complete. I'm on the roll. Thank you, sir.

09:13:53.5 CAPTAIN: Thrust SRS, runway.

09:13:54.7 [Sound similar to increase in engine rpm]

09:14:03.8 FIRST OFFICER: You got throttles.

09:14:08.9 CAPTAIN: Eighty knots, thrust blue. V one. Rotate. V two. V two plus ten.

09:14:30.4 FIRST OFFICER: Positive rate, gear up please.

09:14:31.5 CAPTAIN: Gear up.

09:14:33.1 [Sound of thump and two clicks]

0914:38.5 FIRST OFFICER: Heading select.

0914:41.9 CAPTAIN: Clear left.

09:14:42.6 TOWER: American 587 heavy, turn left. Fly the Bridge Climb. Contact New York departure. Good morning.

09:14:48.3 CAPTAIN: American 587 heavy, so long. Gear's up.

09:14:52.5 FIRST OFFICER: Check speed, level change. Flaps up. Climb power.

09:15:00.0 CAPTAIN: Ah, New York, American 587 heavy, thirteen hundred feet, we're climbing to five thousand.

09:15:04.7 DEPARTURE: American 587 heavy, New

York departure. Radar contact. Climb maintain one three thousand.

09:15:10.2 CAPTAIN: One three thousand feet, American 587 heavy.

09:15:14.6 FIRST OFFICER: One three I see, slats retract.

09:15:16.5 CAPTAIN: Slats.

09:15:17.2 [Sound of several clicks]

09:15:28.5 CAPTAIN: Clean machine.

09:15:28.5 FIRST OFFICER: [Sound similar to yawn] Thank you.

09:15:36.4 DEPARTURE: American 587 heavy, turn left, proceed direct WAVEY.

09:15:37.3 [Sound of brief squeak and a rattle]

09:15:41.0 CAPTAIN: Uh, we'll turn direct WAVEY, American 587 heavy.

09:15:44.4 FIRST OFFICER: Left turn direct WAVEY . . .

09:15:44.7 CAPTAIN: Little wake turbulence, huh?

09:15:45.6 FIRST OFFICER: . . . Yeah.

09:15:47.3 [Sound similar to five sets of stabilizer trim switch clicks]

09:15:48.2 FIRST OFFICER: Two fifty, thank you.

09:15:51.8 [Sound of a thump] [Sound of click] [Sound of two thumps]

09:15:54.2 FIRST OFFICER: Max power. [Spoken in strained voice]

09:15:55.0 CAPTAIN: You all right?

09:15:55.3 FIRST OFFICER: Yeah, I'm fine.

09:15:56.3 CAPTAIN: Hang onto it. Hang onto it.

09:15:56.6 [Sound of snap]

09:15:57.5 FIRST OFFICER: Let's go for power, please.

09:15:57.7 [Sound of loud thump]

09:15:58.5 [Sound of loud bang]

09:16:00.0 [Sound similar to human grunt]

09:16:00.2 [Roaring noise starts and increases in amplitude]

09:16:01.0 FIRST OFFICER: Holy shit.

09:16:01.0 [Sound similar to single ECAM chime.

09:16:02.0 [Sound similar to single ECAM chime]

09:16:04.4 [Sound similar to stall warning repetitive chime for 1.9 seconds]

09:16:06.2 [Roaring noise decreases and ends]

09:16:07.5 FIRST OFFICER: What the hell are we into . . . We're stuck in it [wake turbulence].

09:16:07.5 [Sound similar to continuous repetitive chimes for one second]

09:16:09.6 [Sound similar to continuous repetitive chimes for three seconds]

09:16:12.8 CAPTAIN: Get out of it [wake turbulence], get out of it.

09:16:14.8 END OF RECORDING

The aircraft's vertical stabilizer and rudder separated in flight and were found in Jamaica Bay, about one mile north of the main wreckage site. The aeroplane's engines subsequently separated in flight and were found several blocks north and east of the main wreckage site. All 260 people aboard the aeroplane and five people on the ground were killed, and the aeroplane was destroyed by impact forces and a post-crash fire.

TAOS, New Mexico, USA
8 November 2002

At approximately 1:53 p.m. an Israel Aircraft Industries 1124A transport aeroplane, registered as N61RS, left Las Vegas, Nevada, destined for Taos, where it was expected to pick up a passenger. The aircraft passed the Taos directional beacon at 14:56:39, at an altitude of 15,000 feet. Albuquerque Centre controllers heard a mayday radio call, and radar contact was lost with the aeroplane at 14:57:08, at an altitude of 14,700 feet. There was no further communication with the aeroplane.

An eyewitness, located about a third of a mile north-east of the accident site, reported that he heard 'distressed engine noises overhead', and looked up and saw what appeared to be a small private jet flying overhead. 'The engine seemed to be cutting in and out.' The witness observed the aeroplane 'attempt to bank to the south as it dove down till my view was blocked by the ridge to the west of me'. The witness then heard an explosion and saw a big cloud of smoke rising over the ridge. Another eyewitness, working with cattle about a third of a mile north-west of the accident site, reported that he heard a loud noise and looked up and saw a small white aeroplane with two engines coming from the north-west. The aeroplane 'started to turn left with the nose of the aeroplane slightly pointing toward the ground'. The witness added that the aeroplane appeared to be 'trying to land on the road [US 64]'. The aeroplane appeared to be above the wires when it caught fire. The witness did not see the aeroplane impact the ground; however, moments later there was smoke and an explosion. Then 'a cloud of dust rose up and blew away'.

A third witness, located about one and a half miles west-north-west of the accident site, reported that she heard the roar of the aeroplane's engines. She looked towards the noise and observed the aeroplane in a near vertical descent (nose dive) impact the ground, followed by a 'huge fire ball and puff of smoke'. The witness reported that she 'heard the engines all the way to the ground'.

14:55:29 FIRST OFFICER: Flaps twenty.
14:55:30 [Sounds similar to pitch trim activation tone]
14:55:34 [Sounds similar to pitch trim activation tone]
14:55:36 FIRST OFFICER: Oh, boy.
14:55:47 [Sounds similar to pitch trim activation tone]
14:55:52 FIRST OFFICER: Surprise, surprise, surprise, huh.
14:55:56 [Sounds similar to pitch trim activation tone]
14:55:58 FIRST OFFICER: It's gunna be comin' in.
14:55:59 [Sounds similar to pitch trim activation tone]
14:56:05 CAPTAIN: Gear down.
14:56:08 FIRST OFFICER: Yup.
14:56:09 [Sounds similar to pitch trim activation tone]
14:56:09 [Sounds similar to pitch trim activation tone]
14:56:10 FIRST OFFICER: Here it goes.
14:56:11 [Sound of landing gear being lowered]
14:56:12 [Sounds similar to pitch trim activation tone]

14:56:19 FIRST OFFICER: It's gunna go through.

14:56:21 [Sounds similar to pitch trim activation tone]

14:56:30 FIRST OFFICER: Do I have to shoot a procedure turn?

14:56:33 CAPTAIN: Ah, shit. I don't know. Probably should.

14:56:37 FIRST OFFICER: Wanna make one spin [and] that'll be it?

14:56:38 CAPTAIN: Yeah.

14:56:38 FIRST OFFICER: I don't.

14:56:39 [Sounds similar to pitch trim activation tones in short succession]

14:56:41 CAPTAIN: Shit.

14:56:41 FIRST OFFICER: Oh fuck, what's that.

14:56:42 CAPTAIN: Fuck.

14:56:44 CAPTAIN: Power.

14:56:45 FIRST OFFICER: Full power.

14:56:47 CAPTAIN: Ah, shit [strained voice].

14:56:48 [Sound of wind noise starts to increase until end of recording]

14:56:50 [Sound of strained breathing and grunting starts and continues until end of the recording]

14:56:51 FIRST OFFICER: Come on around.

14:56:52 GPWS: *Bank angle.*

14:56:52 CAPTAIN: Holy shit.

14:56:53 GPWS: *Bank angle.*

14:56:53 FIRST OFFICER: Mayday, mayday, mayday, mayday, mayday.

14:56:53 GPWS: *Caution, terrain terrain terrain.*

14:57:04 AIR TRAFFIC CONTROL: Aircraft calling say again.

14:57:05 FIRST OFFICER: Unload it unload it.
14:57:08 END OF RECORDING

An examination of the accident site revealed that after the aeroplane impacted the ground, aircraft wreckage struck power lines and crossed State Highway 64 at mile marker 230, before coming to rest. The wreckage distribution path was oriented on a measured magnetic heading of 050 degrees. Ground scars indicated that the aeroplane impacted the ground left wing low at a slight nose-down attitude. The furthest piece of wreckage was the right engine, which was located approximately 1125 feet from initial point of impact. The largest piece of the wreckage, a portion of the centre fuselage and fuel cell area, travelled for 925 feet before coming to rest.

The pilot's inadvertent flight into mountain weather conditions resulted in a loss of aircraft control.

CHARLOTTE, North Carolina, USA

8 January 2003

At about 8.47 a.m. Air Midwest Flight 5481, a Beechcraft 1900D, registered as N233YV, took off for Greenville-Spartanburg International Airport, Greer, South Carolina. The aeroplane had been flown from Huntington, West Virginia, the day before with the flight crew reporting that everything was normal.

On the morning of the accident the flight crew was scheduled to fly two flight legs on a one-day-trip sequence. An Air Midwest pilot saw the captain in the gate area at about 7.45 and the first officer at about 8.00. A maximum of thirty-two bags was allowed on the aeroplane. The ramp agent in charge of loading luggage later remembered that two of the checked bags were heavy, with an estimated weight of between 70 and 80 pounds. He told the captain about the heavy bags; the captain indicated that the bags' extra weight would be fine because a child would be on board and would compensate. Air Midwest Flight 5481 departed the gate on time at about 8.30. The captain was the flying pilot. At 08:37:20 the first officer contacted the air traffic control tower ground controller for permission to taxi to runway 18R. At 08:46:18 the tower cleared Flight 5481 for takeoff and instructed the flight crew to turn right to a heading of 230 degrees after takeoff. At 08:46:35 they were cleared for takeoff.

At 08:47:02 the first officer exclaimed, 'Wuh . . .' and the captain said, 'Oh . . .' A second later, the captain said, 'Help me.' At that point, the aeroplane was about 90 feet off the ground and airspeed was 139 knots. At 08:47:04, the captain asked, 'You got it?' Over the next eight seconds, the captain and first officer forcefully tried to push the nose of the aeroplane down. At 08:47:13 the aircraft's pitch attitude of 54 degrees aircraft nose up was at a maximum. The captain radioed air traffic control and declared an emergency. The sound of the stall warning horn ended. At 08:47:18 the aeroplane's pitch attitude decreased through 0 degrees. The aeroplane was about 1150 feet above the ground, with a maximum left roll of 127 degrees and a minimum airspeed of 31 knots. At 08:47:21 the captain stated, 'Pull the power back.' The elevator position reached full aircraft nose up. The stall warning horn started again and continued. At 08:47:26 the aircraft entered a maximum right roll of 68 degrees and a maximum vertical acceleration, and the captain exclaimed, 'Oh, my God, ahhh', and the first officer said something like, 'Uh uh . . . God . . . ahh . . . shit.' The CVR ended at 08:47:28.1.

08:19:08 FIRST OFFICER: How'd you sleep last night?
CAPTAIN: Okay.
FIRST OFFICER: I was just, like, I had one of those aviation nights. I was just like flyin' all night long [in my dreams].
CAPTAIN: Really?

FIRST OFFICER: I was jus' like . . .

CAPTAIN: I had a dream that I was in Miami all night partying.

[Sound of laughter]

08:19:30 FIRST OFFICER: In Miami, huh? Maybe you're destined to go to the DASH . . .

[Sound of laughter]

CAPTAIN: That's what I was thinking.

[Sound of laughter]

FIRST OFFICER: We full?

CAPTAIN: You might want to put the fuel . . . low . . . you know. It's twenty-four. You can count it twenty-two or twenty-three.

FIRST OFFICER: Okay.

08:20:19 CAPTAIN: You can count nineteen people in the back. I don't know bags, yet.

FIRST OFFICER: Okay.

CAPTAIN: Fuel whatever you . . .

FIRST OFFICER: I'll put it at twenty-three . . . I think that's good enough.

08:20:28 CAPTAIN ON INTERCOM: Good morning, welcome aboard USAirways Express service to Greenville-Spartanburg. It's a very short flight, maybe thirty minutes gate to gate. Uh, we ask you to keep your seat belts buckled till [we're at the] gate. [So] anything you brought with you needs to be stowed underneath your seat for takeoff and landing. We have two [emergency exits] on the left, one on the right. To open those doors . . . pull handle down, turn the door sideways, throw the door out and run out. This is your emergency briefing card. It's in the seat in front of you.

We're gonna play a briefing. Please pay attention as we taxi out. This door can also be used as emergency exit. Push the button in the box [and] lift the handle and the door will come out. Please don't hang onto that door. If you do it'll pull you out. Sit back, relax, enjoy the flight and we'll have you there in a few minutes.

08:21:25 CAPTAIN: How many total bags?

RAMP EMPLOYE: Uh, thirty-one.

CAPTAIN: Thirty-one and . . .

RAMP: That's . . .

CAPTAIN: A tyre . . .

RAMP: Including carry-on [baggage].

CAPTAIN: Okay, thank you.

[Crew discusses aircraft weight for two minutes]

08:23:55 CAPTAIN: . . . but we're gonna be okay, right?

FIRST OFFICER: Uh, yeah.

CAPTAIN: Okay.

08:24:02 CAPTAIN: . . . battery on . . . checked annunciator, checked EFIS . . . power's off . . . NAV and beacons on and ground . . . and . . . capped.

FIRST OFFICER: Fuel caps and prop . . .

CAPTAIN: And clear over there?

FIRST OFFICER: Capped and clear on the right.

[Communication with ramp]

08:24:12 CAPTAIN: What?

RAMP: How many we gonna take off?

CAPTAIN: We're figuring it out. We don't think we have to take anything . . .

[Sound similar to engine starting]

08:26:33 CAPTAIN: I didn't know we were gonna be eighteen and overloaded or I wouldn't have . . . let you

. . . I mean, not let you but I . . . not make you do . . . ah
. . . this.

FIRST OFFICER: [Sound of chuckle] No, I got ya.

CAPTAIN: I don't really care how fast or slow you go.

08:26:55 CAPTAIN: Seventeen one twenty is our
weight, huh?

FIRST OFFICER: Yeah, our max . . .

CAPTAIN: So, we're cool.

FIRST OFFICER: So, yeah.

08:27:19 FIRST OFFICER: They show nineteen adult
. . . uh . . .

CAPTAIN: That's okay. You've got . . . Don't worry
about it.

FIRST OFFICER: One child though?

CAPTAIN: Yeah. You can put eighteen [passengers] and
one. You can still count them all as adult; it's your liberty
to do that.

FIRST OFFICER: Oh, really?

CAPTAIN: Yeah.

FIRST OFFICER: I heard other people say if it says
children on there you got to do child weights . . .

CAPTAIN: You just gotta put the child here . . .

FIRST OFFICER: Yeah, yeah. That's what I thought.
That's what I always thought, but . . .

CAPTAIN: You know, it's basically [at the] captain's
discretion. I mean, you're being conservative doing it that
way, so I don't know why . . . You've got Greenville/
Spart . . . Yeah, you got charts up here for Greenville/
Spartanburg . . .

FIRST OFFICER: We're due out at thirty. You wanna
call it?

CAPTAIN: Yeah.

FIRST OFFICER: Thirty? Perfect, yeah. I just gotta plot this index here and then . . . Thank you . . . Uh, seventeen oh eighteen and eighty-one should be in here somewhere.

8:28:21 CAPTAIN: Yeah, just put a circle there. Don't even put a dot. Just make sure it doesn't fall out of the . . . box.

[FIRST OFFICER DEALS FURTHER WITH THE WEIGHT, FINALLY FINISHING]

08:29:02 CAPTAIN: Perfect. Nice job.

FIRST OFFICER: Thank you. Sorry it took me so long.

CAPTAIN: No, it would have taken me just as long. It just seems longer because it's awkward to you . . .

FIRST OFFICER: Yeah.

CAPTAIN: That I'm sitting here doing nothing.

FIRST OFFICER: Yeah, that's true. Good point.

[Laughter]

8:29:22 FIRST OFFICER: [Ramp crewman]'s in a funny mood today.

CAPTAIN: I wonder if I've done something wrong that . . . Is he down there yanking on the chocks?

FIRST OFFICER: Yeah, he just pulled 'em.

[Sound of laughter]

08:29:53 FIRST OFFICER: . . . he's probably lookin' at our . . . tail like [it's] 'bout to hit the ground right now, with all the bags back there.

[Laughter]

CAPTAIN: Yeah.

FIRST OFFICER: Laughing at us.

08:30:23 FIRST OFFICER: And the, uh, after start checklist . . . Avionics master?

CAPTAIN: On.

FIRST OFFICER: Engine anti-ice?

CAPTAIN: On.

FIRST OFFICER: AC buses?

CAPTAIN: On.

FIRST OFFICER: EFIS aux power?

CAPTAIN: On.

FIRST OFFICER: EFIS power?

CAPTAIN: On.

FIRST OFFICER: Load and volt meters are checked, standby attitude indicator is on and uncaged, environmental set, PAX briefing . . . is sent. And the brakes?

CAPTAIN: Released and checked.

PASSENGER BRIEFING RECORDING: Welcome aboard . . .

FIRST OFFICER: Checked right . . .

CAPTAIN: Cleared over there?

FIRST OFFICER: After-start checklist complete. Clear over here.

FIRST OFFICER TO RAMP CONTROL: North ramp Air Midwest 5481 read to taxi out the sewer for Charlie.

RAMP CONTROL: And 5481, north ramp . . . proceed to north ramp hold line, contacting outbound camp one thirty-one point six . . . Be advised there is a pushback top of the concourse.

FIRST OFFICER: To the north hold line . . . Talk to them and we'll look for the pushback at the top, Air Midwest 5481.

08:31:02 FIRST OFFICER: Clear on the right. I still

don't trust these centre – these taxiway . . . I guess they're painted with enough room to – for our right. We got plenty of room over there . . . [but it] never looks like it when you're getting ready to go.

CAPTAIN: I wish we got to fly back from Raleigh to here.

FIRST OFFICER: Yeah, wouldn't that be great . . . Clear on the right. We could tell the crew from the seven-thirty-seven crew that we're just gonna go ahead and fly 'em.

[Sound of laughter]

CAPTAIN: Yeah, good call.

FIRST OFFICER: So that's what I always thought, Katie, that as long as you listed at the bottom, like, you know, on the manifest that you have this many . . . I always thought, like . . . you only use child weights if you absolutely needed to and only then the listing of how many . . . you know, adults, children and infants you had was just in case you ever crashed or something [and] they're looking for bodies, kinda thing.

CAPTAIN: Well, that's what I was under the . . . I mean, I think it is really whatever anybody . . . well, whoever is the captain is the most . . . comfortable with.

FIRST OFFICER: I think you're right. That CJR [a private jet] sure is a good looking plane, isn't it?

CAPTAIN: Yeah. Wish I was flyin' it. [Laughter] I would have to be captain on that because . . . you decide you hate the airlines than you got that Challenger rating already.

08:33:03 FIRST OFFICER: Oh yeah . . . Yeah, wouldn't that be cool?

CAPTAIN: Yeah. We're cleared to the hold line . . .

FIRST OFFICER: Okay. Not to mention you're flyin a sweet piece of equipment.

CAPTAIN: Yeah.

FIRST OFFICER: Yeah, and then like flyin' a Challenger would probably seem like a walk in the park after flyin' that thing . . . You know, like, it's so little.

[Sound of laughter]

08:33:29 CAPTAIN: Yeah, I mean, I might have to, you know, deal with the dinners in Paris or somethin', you know.

FIRST OFFICER: Yeah.

CAPTAIN: Or overnight in Cancun.

FIRST OFFICER: You might suffer through it.

CAPTAIN: Learn another language.

[Sound of laughter]

CAPTAIN: . . . but you know, heck.

FIRST OFFICER: It's just such a better lookin' aeroplane than the ERJs. After not seeing a CRJ back here for a long time I was starting to think, oh, the ERJ is an okay looking plane but they're really kinda ugly when you park a CRJ next to 'em.

08:34:08 CAPTAIN: Who's this? [Seeing another aeroplane land?]

FIRST OFFICER: I don't know . . . Looks like an Airbus 300 maybe? I think?

CAPTAIN: Well, they don't have the winglets on.

FIRST OFFICER: I know, but I always thought they did that to all Airbuses had winglets but like the FedEx. Those planes they have over there, those are all 300s and they don't have winglets either.

214

CAPTAIN: ExpressNet, what's that? That cargo company?

FIRST OFFICER: It's gotta be cargo . . .

CAPTAIN: It's humongous.

[For several minutes while they are making their way to the runway the crew talks about schedules and other airplanes.]

08:35:55 FIRST OFFICER: Taxi checklist. Flight controls free and correct, trims are set, pressurization is set, flight instruments, two niner seven five set and cross-checked.

CAPTAIN: Seven five. Set and cross-checked. We're goin' to spot two, correct?

FIRST OFFICER: Yeah, I agree.

FIRST OFFICER: Aux pumps?

CAPTAIN: Off.

FIRST OFFICER: Autofeather?

CAPTAIN: Armed.

FIRST OFFICER: Ice protection?

CAPTAIN: Standard six.

FIRST OFFICER: Clearance radio takeoff landing data are cleared to Greenville/Spartanburg . . . HUGO five departure . . . radar vectors to Spartanburg. Up to four thousand feet fifty-two twenty-four is a good squawk . . . speeds are four four . . . Twelve and twenty-three. That's reviewed. Crew briefing.

CAPTAIN: . . . [hundred] one twelve one twenty-three. It'll take us out to Greenville/Spartanburg standard brief for the day.

FIRST OFFICER: All right.

CAPTAIN: [Should there] be an engine fire prior to V

one the noting pilot call abort abort [and] the flying pilot abort takeoff . . . Any [other] malfunction . . . my attention [and] I'll [state] abort abort or continue. Any emergency, we will continue run the appropriate items check. Fifteen hundred feet we'll make . . . right traffic one eight right . . . if we have any problems . . . we're at max weight . . . no alternate needed into Greenville/ Spartanburg. One MEL. It's due, it's written on the sixth, due on the seventeenth checklist in the box. You have any questions?

FIRST OFFICER: No questions. Briefs complete, taxi checks complete.

08:37:20 FIRST OFFICER TO CONTROL: Charlotte ground, very good morning, Air Midwest 5481, approaching spot two with Lima.

GROUND: Spot two Air Midwest 5481.

CAPTAIN: Unnn. Gulfstream.

FIRST OFFICER: You just wanna fly a jet. [Sound of laughter]. I don't blame ya.

GROUND: Air Midwest 5481, Charlotte ground. Taxi to runway 18 right.

FIRST OFFICER: To 18 Right, Air Midwest 5481. Good day.

CAPTAIN: That guy doesn't have his fancy voice today.

FIRST OFFICER: See, I thought it was. I thought . . . that this was the guy with the fancy voice, too, so one morning I said, like, how you doing A.D.? 'Cause the guy with the fancy voice told me his name was A.D. or I think that's his first and last initials. Clear on right. Anyway, so one morning I'm talking to this guy and [I

asked him] how're you doin', A.D.? He goes, This isn't
A.D. This is T.K. And so I'm, like, Oh, sorry about that.
It'll never happen again. He's, like, it's okay. A lot of
people confuse us. I've got the ratchet-down seat.
CAPTAIN: Oh, I hate that seat.

[Sound of laughter]

FIRST OFFICER: I still haven't had it – I've heard
those, all those stories about how, like, you'd be flyin'
along and the captain will have the ratchet-down seat
and be flyin' and ILS [landing] or something and all of a
sudden it'd go clunk clunk clunk and it would be, like,
stuck.

CAPTAIN: Ah, you never had that? When you're in
flare, you're like . . . shit . . . wham.

[Sound of laughter]

FIRST OFFICER: . . . I think I'd die laughing 'cause I
heard, like, you know, the captain's just startin' cussin'
up a storm because they're so pissed off, you know, like,
right when, you know, you need the seat to be, like, good
and solid, you go, like, click click click. I've had it when
we're taxiing around, like, we'll just be sitting here and
all of a sudden I'll go three notches down.

[Other conversation about unions]

08:40:40 FIRST OFFICER: Have you seen [the movie]
Caddyshack lately?

CAPTAIN: Huh uh.

FIRST OFFICER: I don't know if you remember or not
but Rodney Dangerfield drives this, like, Bentley or
Rolls-Royce or somethin' that's got this really obnoxious
horn that goes like . . . [makes horn sound] I can't make
the horn sound . . . Anyway, like, when every time he

217

shows up somewhere it's the horn and it plays this whole long thing and I [was] lookin' for it the other day . . . I was bored [and] I went on the internet and I was lookin' for . . . sound files like for Alert sounds, you know, how like you do something wrong your computer bleeps at you? Instead of just like the stupid beep I was lookin' for somethin' a little more fun . . . So, so I found this. I don't know how I came across it but I found a sound clip of that, just that car horn going off. [Sound of laughter] And so . . . And so I, uh, I changed the alert for when I have incoming email on my email programme to that car horn.

CAPTAIN: Oh, God.

FIRST OFFICER: Cracks me up. I start laughing every time 'cause, like, you couldn't possibly miss the email 'cause it plays this long obnoxious car horn sound. It's great.

CAPTAIN: I have that jungle screensaver on my computer.

FIRST OFFICER: Oh, cool.

CAPTAIN: You know if you leave it playin' after a while . . . if you leave the sound on you'll start to hear the birds squawkin' and the monkey yellin' and the . . .

FIRST OFFICER: Nice.

[More discussion of computer sounds]

08:42:31 CAPTAIN: Before takeoff to the line, please.

FIRST OFFICER: Before takeoff to the line, flaps set and indicating seventeen. Props?

CAPTAIN: High.

FIRST OFFICER: Auto-ignition?

CAPTAIN: Armed.

[Checklist complete the crew talks about the mechanical elements of an aircraft they are taxiing past]
08:44:49 FIRST OFFICER: Gosh, that sure is a nice looking plane . . . awe[some]. It's killin' me to see those planes, those CRJs, back here now. It's probably, I should, I should shut up. You're, I'm, I'm doing to you right now what you were doing to me with the Krispy Kremes [donuts] yesterday.

08:45:21 CAPTAIN: Sometimes I secretly go get them in the grocery store.

FIRST OFFICER: Oh, yeah? Yeah, I've been known to do that from time to time.

08:45:25 TOWER: Air Midwest 5481, runway 18 right, taxi into position and hold.

FIRST OFFICER: Position and hold runway 18 right, Air Midwest 5481.

FIRST OFFICER: Uh, below the line, transponders ALT mode and bleeds are set and exterior lights?

CAPTAIN: Are set.

FIRST OFFICER: Before takeoff checklist is complete . . . clear on right. Position and hold.

CAPTAIN: Clear on left. This guy's goin' away.

FIRST OFFICER: Uh, they're going to blast us . . . with his . . . jet blast.

CAPTAIN: I love those Krispy Kreme donuts with the icing filling on the inside.

FIRST OFFICER: You know, I was just sayin', I started to say, ah, right before he told us to taxi in position and hold: you need to get a box of those [donuts] and wrap them up in wedding gift-wrap paper and take them to [someone's] wedding . . .

[Sound of laughter]

CAPTAIN: I'm gonna do that.

FIRST OFFICER: Like a huge box . . . you know like the size . . . that was in the crew room yesterday.

CAPTAIN: That is an awesome idea.

08:46:18 TOWER: Air Midwest 5481, turn right heading two three zero, cleared for takeoff.

08:46:22 FIRST OFFICER: Two three zero, cleared for takeoff . . . Air Midwest 5481. [TO CAPTAIN] Two thirty, cleared to go.

[Sound similar to engine power increasing]

CAPTAIN: Set takeoff power, please.

FIRST OFFICER: Power is set . . . Push. Eighty knots, cross checked. V one, V R, V two, positive rate . . .

CAPTAIN: Gear up.

[Sound similar to landing gear hydraulic motor operating]

08:47:02 FIRST OFFICER: Wuh . . .

[Sound similar to landing gear hydraulic motor ends]

08:47:02 CAPTAIN: Oh.

08:47:03 CAPTAIN: Help me.

08:47:04 CAPTAIN: You got it?

FIRST OFFICER: Oh, shit. Push down. Oh, shit.

08:47:08 [Sound of grunt, exhale]

08:47:10 CAPTAIN: You, uh . . .

[Sound similar to stall warning horn begins]

08:47:11 CAPTAIN: Push the nose down [loud voice] Ahh. [Loud voice] Ahh. Oh, my God.

08:47:16 CAPTAIN TO GROUND: We have an emergency for Air Midwest 5481.

08:47:18 CAPTAIN: [?]: Daddy.

08:47:20 [Sound similar to decrease in engine/propeller noise]
CAPTAIN: Pull back the power.
08:47:26 CAPTAIN: Oh, my God . . . Ahhh.
FIRST OFFICER: Uh uh, God . . . Ahh, shit.
08:47:28 END OF RECORDING

The loss of pitch control during takeoff was the probable cause of this accident. The elevator control system was rigged incorrectly and the aeroplane's aft centre of gravity [perhaps due to the weight and location of the luggage], which was 'substantially aft of the certified aft limit', according to the NTSB, contributed to the destabilizing of the aircraft and the resulting crash. The aeroplane struck a US Airways maintenance hangar on the airport property and came to rest about 1650 feet east of the runway 18R centreline and about 7600 feet beyond the runway 18R threshold. All aboard were killed.

JEFFERSON CITY, Missouri, USA

14 October 2004

On 14 October 2004, Pinnacle Airlines Flight 3701, a Bombardier CL-600–2B19, registered as N8396A, departed Little Rock, Arkansas, to Minneapolis-St Paul at about 9.21 p.m. As this was a repositioning flight, which relocates an aeroplane to the airport where the plane's next flight is scheduled and does not carry passengers or cargo, the only people on board were a captain and first officer. The flight plan included a cruise altitude of 33,000 feet. About six minutes after takeoff, when the aeroplane was at an altitude of about 15,000 feet, the flight crew disengaged the autopilot.

The captain requested a climb to 41,000 feet, which is the aircraft's maximum operating altitude. The first officer said, 'Man, we can do it. Forty-one it.'

At 21:51:51, the first officer stated, 'There's forty-one, oh my man.' Laughing, he added, 'This is . . . great.'

At 21:52:22, the captain asked the first officer whether he wanted a drink and the first officer requested a soda. The captain left his seat shortly afterwards to get the drink. A minute went by, and the captain said, 'Look how high we are.'

Kansas City Air Route Traffic Control Centre asked the pilots whether they were flying a CRJ-200. When the captain confirmed this, the controller stated, 'I've never seen you guys up at forty-one there.' The captain replied, 'We don't have any passengers on board so we decided to have a little fun and come on up here.' He added, 'This is actually our service ceiling.'

The crew was clearly taking a joyride. Some Pinnacle Airlines pilots had expressed curiosity about operating the aeroplane at 41,000 feet and that an informal 'FL [flight level] 410 [41,000 feet] Club' existed at the airline.

At 21:54:07 the captain told the first officer, 'We're losing here. We're gonna be . . . coming down in a second here.' Three seconds later, the captain stated, 'This thing ain't gonna . . . hold altitude, is it?' The first officer responded, 'It can't, man. We greased up here but it won't stay.' The captain reiterated, 'Yeah that's funny. We got up here it won't stay up here.' He contacted the controller and said, 'It looks like we're not even going to be able to stay up here . . . look for maybe . . . three nine oh or three seven.' The aeroplane's airspeed had decreased to 150 knots. The stick shaker and the stick pusher activated three times, before the left and right engines' fan speed and fuel flow began decreasing.

At about 9.54 both engines flamed out. The captain told the ground controller, 'Declaring emergency. Stand by.' For the next fourteen seconds the flight crew made several control column, control wheel and rudder inputs and recovered the aeroplane from the upset at an altitude of 34,000 feet.

At 21:55:14 the ground controller told the pilots to descend and maintain an altitude of 24,000 feet; about five seconds later, the captain acknowledged the assigned altitude. At 21:55:23, one pilot stated to the other, 'We don't have any engines', and about ten seconds later the captain stated, 'Double engine failure.' The crew tried in vain to restart the engines. At 22:14:02 the first officer told the captain that he had the runway in sight. The captain questioned the first officer about the location of the runway and then stated, at 22:14:17, 'We're not gonna make this.' He asked, 'Is there a road? We're not gonna make this runway.' The aeroplane then turned left and headed towards a straight and lit section of highway. At 22:14:46, the captain stated, 'Let's keep the gear up . . . I don't want to go into houses here.'

21:49:07 CAPTAIN: Forty thousand, baby.
FIRST OFFICER: Come on.
CAPTAIN: Look at that cabin altitude, man.
   [Sound of laughing]
   [Sound of altitude alert]
CAPTAIN: Thousand to go. Should be at eight thousand feet moving slowly [going up].
CONTROL: Flagship 3701, would you like to go direct to KASPR?
   [Sound of laughter]
CAPTAIN TO FIRST OFFICER: Sure.
FIRST OFFICER: Might as well.
FIRST OFFICER TO CONTROL: Yeah, that'll be great, 3701 direct KASPR.
CONTROL: All right, cleared direct KASPR, Flagship 3701.
FIRST OFFICER: Thank you, sir. Appreciate that, 3701 going direct KASPR. You saved us two minutes.

21:49:49 [Sound of laughter]

CAPTAIN: That's fucking crazy.

[Sound of laughter]

21:50:30 [Sound of whistling]

CAPTAIN: I'm saying, don't let it get below 170. We're levelling off here anyway, so . . .

FIRST OFFICER: Dang. [Sound of laughing] Our arrival fuel's supposed to be 3.5.

CAPTAIN: I can't believe that shit, man. That's crazy. We've saved a ton of fuel [by climbing to this altitude].

[Sound of laughing]

CAPTAIN: That's what I mean. I'll leave the power up 'til we get [to level off].

FIRST OFFICER: We're at V-T . . .

21:51:51 FIRST OFFICER: There's forty-one [41,000 feet]. Oh, my man . . . Made it, man.

CAPTAIN: Yeah.

21:51:58 CONTROL: Flagship 3701, contact Kansas City one two five point six seven.

CAPTAIN: Twenty-five sixty-seven. You have a good night, 3701.

FIRST OFFICER: This is great.

CAPTAIN: Kansas City Centre, good evening. Flagship 3701, four one [41,000 feet].

CONTROL: Flagship 3701, Kansas City Centre, roger.

[Sound of laughter]

CAPTAIN: You'll get the . . . you'll do the next one to say four one . . . oh, yeah, baby.

[Laughter]

21:52:22 FIRST OFFICER: Four one . . . oh, four one . . .

CAPTAIN: Want anything to drink?

FIRST OFFICER: Ah, yeah, actually. I'll take a Pepsi.

21:52:30 CAPTAIN: A Pepsi? I thought you said, 'Beer, man.' Yeah, I'd like one, too.

FIRST OFFICER: Is that seal on the liquor cabinet?

[Laughter]

21:53:14 FIRST OFFICER: This is the greatest thing. No way . . .

CAPTAIN: You want a can? You want a cup? We don't have any ice . . .

FIRST OFFICER: That's fine.

CAPTAIN: They're cold as hell, dude.

21:53:24 CAPTAIN: Accelerating up at all?

FIRST OFFICER: [Laughter] No, man.

CAPTAIN: Nothing, dude.

FIRST OFFICER: It ain't speeding up with shit.

CAPTAIN: Look how high we are.

   [Laughter]

CAPTAIN: That nose is . . . Look at how nose-high we are.

FIRST OFFICER: I know . . . Dude, that fuckin' ball's way off. Dude, the ball's full off.

CAPTAIN: No shit. Look at this ball, dude.

21:53:42 CONTROL: Flagship 3701, are you a RJ 200?

CAPTAIN: 3701, that's affirmative.

FIRST OFFICER: Forty-four hundred [44,000 feet].

CONTROL: I've never seen you guys [RJ 200 Aircraft] up at forty-one [41,000 feet] before . . .

   [Sound of laughter]

CAPTAIN: Yeah, we're actually . . . ah . . . We don't have any passengers on board so we decided to have a little fun and come on up here.

CONTROL: Gotcha.

CAPTAIN: This [is] our . . . actually our service ceiling. [TO FIRST OFFICER]: Damn thing's losing it.

[Laughter]

CAPTAIN: We're losing here . . . We're gonna be coming down in a second here, dude. This thing isn't gonna hold altitude. Is it?

FIRST OFFICER: It can't, man. We greased up here but it won't stay.

CAPTAIN: Yeah, that's funny. We got up here but [it] won't stay up here.

FIRST OFFICER: Dude, it's losing it.

[Laughter]

CAPTAIN: Yeah.

CAPTAIN TO CONTROL: And centre, 3701.

CONTROL: Go ahead.

CAPTAIN: Yeah, just as you said. It looks like we're not even going to be able to stay up here. Ah, look for maybe . . . ah, three nine oh or three seven [39,000 or 37,000 feet].

CONTROL: Flagship 3701, stand by.

[Sound of stick shaker] [Sound of autopilot disconnect]

21:54:43 FIRST OFFICER: What'd he say?

CAPTAIN: I dunno.

CONTROL: Say again, for 3701.

[Sound of stick shaker]

CAPTAIN: Shit.

FIRST OFFICER: I got it.

21:54:56 CAPTAIN: Shit.

[Recorded sound] *Engine oil.*

CAPTAIN: Come on, come on. [Engines flame out]

CONTROL: Flagship 3701, say . . .

CAPTAIN: Declaring emergency . . . Stand by . . . Aw, shit.

[Sound of triple chime, similar to master warning alert]

[Recorded sound] *Engine oil. Engine oil. Engine oil.*

CONTROL: Descend at pilot's discretion . . . maintain . . . Flight level two four zero [24,000 feet]

CAPTAIN: Two four zero . . . flight . . . two four zero. The important thing is . . . We don't have any engines.

[Engine oil alert]

21:55:31 CAPTAIN: Pull . . . pull the handle.

CONTROL: Everybody stand by . . . Flagship 3701, the frequency's open . . .

CAPTAIN: Shit . . . Got the aeroplane? [TO FIRST OFFICER] You gotta be kidding me.

FIRST OFFICER: All right, ah . . . [Stand by for map light and] dome.

CAPTAIN: We're still descending, aren't we? Are we holding altitude?

FIRST OFFICER: Ah, yeah. We got it.

CAPTAIN: Okay.

FIRST OFFICER: We got a little bit of engine [windmill] in one of them [one engine].

CAPTAIN: Really? Okay, we gotta go to emergency . . .

FIRST OFFICER: We're not hold[ing] altitude.

CAPTAIN: We're not?

FIRST OFFICER: No, we're not.

CAPTAIN: Ah, flashlight [Where is the flashlight], dude?

FIRST OFFICER: Flashlight's in my bag, my bag.

CAPTAIN: There's bags all [over the place]. Look back here. Double engine failure. You holding altitude?

FIRST OFFICER: Ah, no I'm not.

CAPTAIN: Okay, continuous ignition on. Thrust levers shut off . . . Restart, shut off . . . Power's established. How do you know power's established?

[They try to restart the engines]

[Sound of triple chime, similar to master warning alert]

[Recorded sound] *Cabin pressure.*

CAPTAIN: Stab trim channel two. Engaged. Target airspeed established above flight level three four oh [34,000 feet]. We're below. Point seven Mach. So look for point seven Mach. A hundred eighty . . . ADG below thirty thousand feet . . . Okay, descend below thirty thousand feet.

[Recorded sound] *Cabin pressure. Cabin pressure. Cabin pressure.*

FIRST OFFICER: Okay.

CAPTAIN: Shit, dude.

FIRST OFFICER: Damn gear's unsafe . . .

CAPTAIN: I'll worry about that later.

21:57:54 CAPTAIN: Is the gear down or . . . unsafe? Okay, go, go. Descend still.

FIRST OFFICER: All right.

CAPTAIN: You got a question on [why/what?].

FIRST OFFICER: That was a Dutch roll, I believe.

[Dutch roll: a combination of 'tail-wagging' and rocking from side to side]

CAPTAIN: It was putting and pushing.

FIRST OFFICER: Sure.

CAPTAIN: See the plane start to roll on us.

FIRST OFFICER: We were descending at two thousand feet per minute. Is the gear down . . . unsafe?

CAPTAIN: Okay, go. Go. Descend still.

FIRST OFFICER: All right. We need our oxygen masks.

CAPTAIN: Okay, as soon as we're above . . . below thirty thousand we can start the APU [auxiliary power unit].

21:58:37 FIRST OFFICER: Go on oxygen?

CAPTAIN: You know what. Yeah, we need to go on oxygen.

[Sound of oxygen flow starting in oxygen mask]

[Sound of chime similar to master caution alert]

21:59:25 CAPTAIN: Okay, we have power. Slow it away.

CONTROL: . . . and centre, Flagship 3701. . . Flagship 3701, go American 751 . . . Stand by.

CAPTAIN: Yeah, we're still descending. We're gonna need to descend down . . . ah, probably lower . . . probably gonna descend down to . . . right now . . . to about thirty thousand feet. Is that okay?

CONTROL: Flagship 3701, affirmative. Descend and maintain one three thousand [13,000]. Your local altimeter setting is . . . ah, oh stand by. Two niner six five . . . and ah, one three thousand [13,000 feet] is approved, Flagship 3701.

CAPTAIN: All right, two nine six five, thirty-seven zero one [37,000 feet].

22:00:22 FIRST OFFICER: No, no, no, what do you got . . . oh, heading.

CAPTAIN: Yeah.

FIRST OFFICER: See what you got.

CAPTAIN: Yeah, yeah, it's two nine six five on altimeter setting.

FIRST OFFICER: Okay.

CAPTAIN: When we come through [sound of page turning] we're gonna have to descend down to . . . ah . . . [page turning as captain looks in manual] . . . thir . . . thirteen [thousand feet], okay? Actually, push the nose over . . . Push it over. Let's get above three hundred knots.

FIRST OFFICER: Okay.

CAPTAIN: Twenty-one thousand feet . . . We need, we need . . . Check our airspeed and altitude.

FIRST OFFICER: Three hundred.

CONTROL: Flagship 3701, are you able to take a frequency change at this point?

CAPTAIN: 3701, stand by.

CONTROL: Roger.

CAPTAIN: . . . 'kay, it's getting to it. Yeah, it's coming down now. I don't think we're gonna need that. Ignition on.

FIRST OFFICER: Yep.

CAPTAIN: [Reading?] Airspeed not less than three hundred knots . . . You wanna push it up there three hundred knots. Altitude loss approximately can be expected from two hundred forty to three hundred knots. I-T-T ninety degrees or less. N-2 is at least twelve per cent . . . N-2 . . . No, we're not getting any N-2 at all. So we're gonna have to . . . gonna have to go to here . . . thirteen thousand feet. We gotta go down here, dude. We're going to use the APU bleed air procedures [to start engines].

[Sound of chime, similar to master caution alert]

22:02:09: CAPTAIN: Oh, shit. We need to slow it

down. Slow the rate of descent down . . . Target airspeed is established. Target airspeed is hundred ninety knots . . . hundred seventy knots. Go ahead and pull back to a hundred seventy knots. Left and right tenth stage . . . tenth stage bleeds closed . . . left and right tenth stage bleeds closed. APU load control valve open . . . Continuous ignition.

[Sound of chime, similar to master caution alert]
22:02:34 CAPTAIN: No, keep us going down . . . Oh, you know what? Actually we can't do that yet.
CAPTAIN TO CONTROL: And 3701, we can change frequency at this time.
CONTROL: Flagship 3701, ah, roger. I'll have a frequency change for you in just a moment . . . Before I send . . . what was the nature of your emergency, please?
CAPTAIN: Ah, we had an engine failure up there at altitude . . . It . . . ah, aeroplane went into a stall and one of our engine's failure . . . So we're gonna descend down to start our other engine.
CONTROL: Okay, that's kinda what we were figuring there, and ah, understand you're controlled flight and, ah, you're gonna be able to return to normal when you get to lower altitude?
CAPTAIN: Right now, we're not . . . we're stand by for that . . . We're descending down to thirteen thousand to start this other engine. We'll tell you . . .
CONTROL: Flagship 3701, roger. Understand controlled flight on a single engine right now and, ah, I'll go ahead and relay that. You can contact Kansas City Centre on one three four point five. Just advise, ah, her of your intentions. One three point five. Good day.

CAPTAIN: Thirty four five, good day. [Switches frequency] Centre, Flagship 3701, with you, ah, coming through eighteen thousand for thirteen.

CONTROL: Flagship, 3701, Kansas City Centre, roger. And advise of any further help you might need.

CAPTAIN: Will do, 3701. [TO FIRST OFFICER] Okay, get on oxygen.

FIRST OFFICER: Yeah, get on oxygen, dude. We're at cabin altitude. I got it. Fifteen thousand four hundred. We need to be on oxygen.

CAPTAIN: Okay, it's gonna be from thirteen thousand feet and below target airspeed established. It's a hundred and seventy knots. Left and right tenth bleed will be closed. APU open, continuous ignition check. It's on. Continuous ignition is on. Left or right engine start. Let's start number two first . . . Push thrust lever at idle. 'Kay, yeah, these are off right now. [TO CONTROL]: And centre, 3701, we're gonna need a little lower [altitude] to start this other engine up, so we're gonna go down to about twelve or eleven [thousand feet], is that cool?

CONTROL: Flagship 3701, affirmative descend and maintain . . . you wanna go down to eleven or twelve?

CAPTAIN: Ah, we'll go down to at least eleven thousand, 3701.

CONTROL: Flagship 3701, descend and maintain one one thousand and just advise you want to go back to KASPR? Do you want to land? What do you want to do?

CAPTAIN: Ah, just stand by. Right now, we're gonna start this other engine and see . . . make sure everything's okay.

CONTROL: [You] have a lot of choices up ahead.

Columbia's right up ahead. JEF's up ahead and they're the best to accommodate you.

CAPTAIN: Roger, 3701. Thank you. [CAPTAIN TO FIRST OFFICER] Okay, thirteen thousand feet. It says . . . right left tenth stage closed. They're closed. [APO's/APU isolation's] valve open. It's open, dude. Let's check . . . ready to start. Here goes number one [engine]. Start . . . Time started.

22:07:59 CAPTAIN: It's starting. Right engine start. We're not getting any N-2. Aw, shit. Off oxygen. Um, switch.

FIRST OFFICER: Yeah . . .

CAPTAIN: Put it over there. Hold this. It's still on.

FIRST OFFICER: You got it?

CAPTAIN: I got it. Pull that checklist up.

FIRST OFFICER: Okay.

CAPTAIN: Tell her. Tell her we need to get direct to airport. Neither engine's started right now.

CONTROL: Flagship 3701, what altitude do you want to go down to?

FIRST OFFICER TO CONTROL: 3701, we need direct to any airport. We have a double engine failure.

CONTROL: All right. You want to go direct to JEF?

CAPTAIN: Any airport and closest airport.

FIRST OFFICER TO CONTROL: Closest airport. We're descending fifteen hundred feet per minute. We have nine thousand five hundred feet left.

CONTROL: Flagship 3701, cleared direct JEF.

CAPTAIN: Okay, let me see . . . start both engines. They're closed. Tenth stage closed. APU is on. Getting power?

234

FIRST OFFICER: No.

22:09:54 CAPTAIN: Let's try this.

FIRST OFFICER: Fuck.

CAPTAIN: [Naw, dead] Fuck this shit. Where do we have to go?

FIRST OFFICER: JEF. Right in front of you fifteen miles.

CONTROL: Flagship 3701, descend at pilot's discretion [and] maintain three thousand. They're landing ILS runway 30 and . . .

CAPTAIN: Get a frequency, gotta get a frequency.

CONTROL: The wind's are two niner zero at six knots.

FIRST OFFICER: ILS three zero. What is the frequency, please?

CONTROL: Let me give you the frequency for Mizzo [Missouri] approach is one two four point one.

FIRST OFFICER: The approach frequency is one two four one or what is the ILS frequency?

CONTROL: Let me get you the ILS frequency.

FIRST OFFICER: What is the ILS frequency gain?

CAPTAIN: Ask her . . . We're in the middle of the fucking dark here.

FIRST OFFICER: Yeah, we're running.

CAPTAIN: You get her on the radio? Talk to her.

CONTROL: It's one one zero point five.

FIRST OFFICER: Thank you much.

CAPTAIN: One one zero five.

CONTROL: Flagship, did you get that? One one zero point five.

FIRST OFFICER: One one zero point five. [TO CAPTAIN] Were going to green needles?

CAPTAIN: Yeah.

22:11:37 FIRST OFFICER: Okay, should we try starting her up?

CAPTAIN: Yeah, yeah. You might as well. Try it, dude.

FIRST OFFICER: I dunno if this thing is still starting.

CONTROL: Flagship 3701, MIA is twenty-seven hundred.

FIRST OFFICER: Roger, thanks.

CAPTAIN: What is MIA?

FIRST OFFICER: I don't know, man.

22:12:05 FIRST OFFICER: Why isn't the fucking engine going anywhere?

CAPTAIN: I dunno. We're not getting any N-2.

FIRST OFFICER: We're not?

CAPTAIN: Left engine oil pressure. For some reason it's shut down. I don't get it either. Ask her [CONTROL] how we look.

FIRST OFFICER: 3701, how do we look for the airport?

CONTROL: Okay, the airport is at your twelve o'clock and okay, let's . . . let's make that eleven o'clock and eight miles.

CAPTAIN: How do we look for the runway?

CONTROL: Okay. From you it is a three sixty heading.

CAPTAIN: Three sixty heading . . . Turn in now?

CONTROL: Flagship 3701, three sixty heading eight miles.

CAPTAIN: We're turning left. Left turn three sixty heading. [TO CONTROL] Are we gonna make this airport? We don't have the airport in sight. We're heading three six zero now. Do you have any further information? [TO FIRST OFFICER] You try yours. I'm not getting through to her [CONTROL].

FIRST OFFICER: How do we look now, three six heading. We do not have the airport in sight.

CONTROL: And keep turning left. It's now about a three fifty heading.

FIRST OFFICER: Turning three fifty. I have the beacon in sight.

CAPTAIN: Where? Where?

FIRST OFFICER: Twelve o'clock. Straight ahead.

CAPTAIN: Straight ahead? Where's the runway? Are we lined up for the runway?

FIRST OFFICER TO CONTROL: I do not see the runway. I have the beacon. Where's the runway?

CAPTAIN: Come on, lady. Talk to her again.

FIRST OFFICER: Flagship 3701, have the beacon twelve o'clock. The runway is at heading zero three zero?

CONTROL: Flagship 3701, the beacon is on the far side of the runway.

FIRST OFFICER: Okay, I think I have the approach end in . . . [sight]. Here it is at twelve o'clock right.

CAPTAIN: Where?

FIRST OFFICER: Straight ahead.

CAPTAIN: Straight ahead. We're on the approach?

FIRST OFFICER: Yes. Just turn to the right a little bit.

CAPTAIN: Turn to the right a little bit?

FIRST OFFICER: Stay right there.

CAPTAIN: Right here?

FIRST OFFICER: Yeah.

22:14:17 CAPTAIN: Dude, we're not gonna make this thing.

FIRST OFFICER: [I] think we're okay.

CAPTAIN: Where is it? I don't know.

22:14:36 FIRST OFFICER: We're not gonna make it, man. We're not gonna make it.

22:14:38 CAPTAIN: Is there a road? Tell her we're not gonna make this runway.

FIRST OFFICER TO CONTROL: We're not gonna make the runway. Is there a road?

   [Recording: *Too low gear*]

22:14:46 CAPTAIN: Let's keep the gear up. I don't want to go into houses here.

FIRST OFFICER: [There's a] road right there.

22:14:52 CAPTAIN: Where?

FIRST OFFICER: Turn turn.

CAPTAIN: Turn where?

FIRST OFFICER: Turn to your left, to your left.

   [Recording: *Too low terrain*]

22:14:59 CAPTAIN: Can't make it.

   [Recording: *Whoop, Whoop, Pull up*]

22:15:03 CAPTAIN: Aw, shit. We're gonna hit houses, dude.

   [Recording: *Whoop, Whoop, Pull up*]

22:15:06 [Sound similar to impact]

22:15:07 END OF RECORDING

After an investigation, the NTSB determined that the probable causes of this accident were (1) the pilots' unprofessional behaviour, deviation from standard operating procedures and poor airmanship, which resulted in an inflight emergency from which they were unable to recover, in part because of the pilots' inadequate training; (2) the pilots' failure to prepare for an emergency landing in a timely manner, including communicating with air traffic controllers immediately after the emergency about the loss of both engines and the availability of landing sites; and (3) the pilots' improper management of the double engine failure checklist, which allowed the engine cores to stop rotating and resulted in the core lock engine condition.

The captain and first officer both died. The aircraft was destroyed.

CHICAGO, Illinois, USA
8 December 2005

Southwest Airlines flight 1248, a Boeing 737–7H4, registered as N471WN, departed from Baltimore/Washington
Airport at 5.58 p.m. for Chicago Midway Airport. The
flight went as planned. The landing didn't.

FIRST OFFICER: And folks good evenin' ah again
from the guys up front thank you so much for your
patience this evenin' with our troubles with mother
nature we sure do appreciate your understanding . . .
[We] are at thirty-nine thousand feet [and] we're gonna
make a slow climb to forty thousand feet here in just a
couple minutes. Our friends at air traffic control [tell] us
it's gonna be a pretty smooth flight this evenin'. [Sound
of seat belt warning tone] The seat belt sign's comin' off
so folks if you wanna get up and stretch your legs or
move about you sure can do so. Well please folks while
you're in your seats always stay buckled up just in case
we do find a few unexpected bumps along the way. Folks
as long as the flight attendants are busy trying to get their
service out, you sure can do them a big favour by trying
to keep those aisles clear. They really do appreciate it . . .
You can't stand up in the forward galley area. These are
FAA security regulations and each of us want to thank

240

you here in advance for complyin' . . . Strong headwinds right now folks are out of the west at about a hundred and fifty knots 'bout a hundred seventy-five mile an hour right on the nose. Normal, ah, normally our speed across the ground is close to about five hundred miles an hour. Right now it's only about three hundred and fifty or so. It's not gonna help us make up any time at all right now [with] about four hundred and fifty miles to go . . . Showin' us, ah, – touchin' down in Chicago [in] about an hour twenty minutes or so 'bout [in] an hour twenty-five minutes we should have you safely at the gate. Folks as we get closer we'll keep you posted, get you up to date on the latest for the Chicago weather. For now, folks, again, ah, thanks so much for your patience and understanding this evening. We really do appreciate it and finally folks one extra note. We have three great flight attendants and I promise ya they didn't have a single thing to do with us gettin' off to the late start this evenin'. So folks please be kind to them 'cause they're gonna take excellent care of you. Thanks so much folks again and good evenin'.

17:23:51.7 CAPTAIN: Wow whaddya say?

17:23:53.8 FIRST OFFICER: Yadda yadda yadda. yadda yadda yadda. Good flight attendants yadda yadda. We like visitors yadda yadda.

17:24:01.6 CAPTAIN: Wow.

17:24:03.7 FIRST OFFICER: I always give 'em the be-good-to-our-flight-attendants-'cause-it's-not-their-fault-we're-late. They're, they always say ohh that's so nice, thank you.

17:24:13.9 CAPTAIN: Huh.

17:24:14.2 FIRST OFFICER: It's all lies and garbage.

17:24:15.5 CAPTAIN: Yep.

17:24:17.0 FIRST OFFICER: Windy windy windy.

17:25:01.0 CAPTAIN: Big three hundred and thirteen knots groundspeed.

17:25:04.2 [A full minute of chatter goes by]

18:03:22.3 FIRST OFFICER: You working Christmas?

18:03:23.9 CAPTAIN: Say again?

18:03:24.5 FIRST OFFICER: You working Christmas?

18:03:25.5 CAPTAIN: I am working Christmas.

18:03:26.8 FIRST OFFICER: Be where ya gotta be.

18:03:28.5 CAPTAIN: I picked it up for a guy so he could be home with his three-year-old and six-year-old.

18:03:33.3 FIRST OFFICER: Aren't you a nice guy.

18:03:35.9 CAPTAIN: Where ya gonna be?

18:03:37.5 FIRST OFFICER: I just do a turn.

18:03:39.0 CAPTAIN: Oh cool.

18:03:39.9 CAPTAIN: Back through Baltimore unfortunately. There's nothin' there.

18:03:43.4 FIRST OFFICER: Yeah. You're a good man.

18:03:50.8 CAPTAIN: Hmmm.

18:04:13.8 FIRST OFFICER: Do you wanna let the girls up again?

18:04:15.4 CAPTIAN: Oh, yeah. Yeah, go ahead.

18:04:18.6 FIRST OFFICER: I'll be right back. Let everybody up?

18:04:21.8 CAPTAIN: Sure.

FIRST OFFICER: Hey folks and from the guys up front you have a few minutes if you need to get up again, ah, you sure can do so . . . [We'll be coming] into Chicago in, ah, about another thirteen minutes and that's when I'll turn the seat belt sign back on . . . A total distance two hundred twelve miles looks good to be in the gate, ah, touchin' down at fifty after in the gate about fifty five past. Thanks so much folks.

18:06:13.4 CAPTAIN: Still snowin', isn't it?

18:06:14.5 FIRST OFFICER: Yeah, freezin' fog.

18:07:35.6 CAPTAIN: I don't know if I'm comfortable usin' the autobrakes . . .

18:07:39.2 FIRST OFFICER: Yeah.

18:07:39.5 CAPTAIN: In, ah, in this, ah, situation.

18:07:43.4 FIRST OFFICER: First time.

18:07:44.3 CAPTAIN: You know havin' not even seen 'em operate before and then all of a sudden go in with a . . .

18:07:47.9 CONTROL: Southwest 1248, descend and maintain flight level 36,000.

18:08:10.7 FIRST OFFICER TO CONTROL: Flight level three six zero Southwest 1248. You have any complaints about the rides in the descent?

18:08:14.9 CONTROL: I think what's – in this area anyway it's, ah, twenty [thousand feet] and below are the only bad rides.

18:08:21.1 FIRST OFFICER TO CONTROL: Okay thanks a lot.

[A further ten minutes go by.]

CONTROL: Southwest 1248, turn right heading three five zero for sequence [and] maintain two hundred and fifty knots. Well, what are you doing right now actually for knots?

18:18:23.2 CAPTAIN TO CONTROL: Two ten.

18:18:23.8 FIRST OFFICER TO CONTROL: We're

slowed to two ten we're heading right to three fifty for Southwest 1248.

18:18:27.6 CONTROL: And Southwest 1248 you can keep your current airspeed [and] fly heading three five zero. They're out of the hold. They gotta sequence you now.

18:18:33.0 FIRST OFFICER TO CONTROL: Okay, great. Three fifty heading and two ten [speed], Southwest 1248.

18:18:38.4 CAPTAIN: Two fifty heading.

18:19:36.4 FIRST OFFICER TO CABIN: [We're] startin' our very gradual descent into Chicago Midway Airport [sound similar to single chime]. Ah, seat belt sign's comin' back on, so folks please if you're up and about head on back to your seats . . . for the duration. There's only a hundred and twenty-three miles between us and the airport. That's the good news . . . get our sequence into arrival . . . into Chicago. . . ah, out of the east at ten miles an hour, low visibility due to on-and-off light snow and it's very chilly – it's only twenty-five degrees. Folks again, ah . . . this evenin' we sure do appreciate all your patience and understanding and all and we welcome you to Chicago. Folks, if you're continuing [to] Las Vegas and then finally to Salt Lake City with us, we're gonna be on the ground in Chicago hopefully for only about twenty-five minutes and we're gonna get you [safely on your way]. Yeah, folks, thanks so much for your patience and understanding tonight [and we hope] everybody has a wonderful night on your way home. Buckle up, drive safely and next time you're gonna go flying, we'd sure love for you to come back and see us again here at Southwest. Thanks so much, folks. Good night.

18:20:23.8 CONTROL: Southwest 1248, turn left heading three one zero.

18:20:27.6 CAPTAIN: Three one zero Southwest 1248

18:25:44.9 CAPTAIN: Oh three thousand feet.

18:25:46.7 FIRST OFFICER: Snow increasing.

18:25:58.6 CONTROL: . . . three Kilo F . . . they are ploughing the runway at Midway now, so the next sector is in the hold. You can expect to hold on the next frequency reaching 10,000 slow to two hundred and fifty knots.

[CAPTAIN AND FIRST OFFICER CHAT ABOUT THE *SESAME STREET* CHARACTER ELMO]

18:30:22.5 FIRST OFFICER: How many kids did you have?

18:30:23.7 CAPTAIN: I have three, ah, – yeah, three children and two grandchildren.

18:30:37.5 FIRST OFFICER: Two grandkids? Any more grandkids on the horizon?

18:30:39.9 CAPTAIN: Nah, not yet.

18:30:44.9 FIRST OFFICER: That's cool.

18:30:45.4 CAPTAIN: And he's been married a little over a year so they're thinkin' about it. [They] weren't thinkin' about it before but they are now.

18:30:52.1 FIRST OFFICER: I never told the girls [flight attendants] to sit either.

18:30:54.6 CAPTAIN: Well . . . I know I haven't told 'em to clean up yet.

CAPTAIN TO FLIGHT ATTENDANTS: Ah, flight attendants take your seat if you haven't already. Just keep your seats till we let ya up, thanks.

245

18:36:32.6 CONTROL: Ah, right now my understanding is they're just clearin' the runway [of snow] and I should get an update here shortly.

CAPTAIN TO CONTROL: [Can you] monitor the – I'm gonna I'll talk to the folks real quick.

18:37:24.0 CAPTAIN ON PA SYSTEM TO CABIN: Well, folks, you've probably felt that we've slowed here. We're gonna have to hold here for about twenty minutes. It looks like till they clear the runway off. They've been off and on doing that as we've been flying into Chicago to get the braking action such that we can land. So it's still snowing so the ploughs are workin' on the runway. We're gonna hold and let 'em do their job and then once they get . . . then we'll be heading into Chicago. So they gave us a fifty-five after the hour expected approach clearance time and that's when we expect to leave the holding pattern and head to (Chicago) . . .

18:38:12.3 FLIGHT ATTENDANT TO CABIN: Ladies and gentlemen, we are [about to make our] final approach into Chicago . . . Make sure your seat belts are fastened . . . lean forward and press the silver button . . .

18:50:25.8 FIRST OFFICER TO CAPTAIN: What ya do before? Where did you fly before?

18:50:29.0 CAPTAIN: What'd I fly before?

18:50:30.0 FIRST OFFICER: Yeah.

18:50:30.4 CAPTAIN: I was an Air Force guy.

18:50:31.6 FIRST OFFICER: How long [ago], did you retire?

18:50:32.9 CAPTAIN: I was, I been here eleven years so . . .

18:50:36.7 FIRST OFFICER: How long did you stay in the Air Force for?

18:50:43.6 CAPTAIN: Twenty-six.

18:50:44.4 FIRST OFFICER: Twenty-six years? How old are you?

18:50:46.8 CAPTAIN: I'm fifty-nine.

18:50:47.9 FIRST OFFICER: Are you serious?

18:50:48.5 CAPTAIN: Yeah.

18:50:49.2 FIRST OFFICER: Jeeze, you'd never freakin' know it. You gonna make it? It's [the airlines' mandatory retirement age] gonna change?

18:50:52.8 CAPTAIN: I don't know if they cha . . . I . . . a November . . . I've got till November eight [before the airlines change the age of retirement or he will have to retire].

18:50:58.5 FIRST OFFICER: I hope it does. Some of my favourite people had to go [and retire] and it was too bad . . . he was . . . awesome guy to fly with outta Chicago. I flew with [him on] his second to last night.

18:51:11.9 CAPTAIN: Is that right?

18:51:22.8 FIRST OFFICER: Fifty-nine man. You'd never know it.

18:51:25.7 CAPTAIN: Ha, tell me about it.

18:51:52.1 FIRST OFFICER: I'm gonna give 'em a quick update since we told 'em fifty.

18:51:56.4 FIRST OFFICER ON PA SYSTEM: Folks, just to give you a fast update. We are estimating one more turn in the hold. They're accepting aircraft back in Chicago Midway right now. I think we're about three or four in the stack so hopefully we'll be heading that way soon. We'll get back to ya but in the meantime, folks, thanks so much for your patience and hopefully we'll have you headin' towards the airport here in just a couple a minutes. Thanks . . . we're headin' towards the airport [with] forty-seven miles to go. Should be in the gate by about ten after. Thanks so much. And again everybody have a great night. Thanks so much for your patience.

18:57:34.1 CONTROL: Southwest 1248, turn right heading zero five zero.

18:57:37.5 FIRST OFFICER: Zero five zero for Southwest 1248.

18:57:39.5 CAPTAIN: Zero five zero.

19:01:19.3 FIRST OFFICER: Six thousand Southwest 1248.

19:02:27.1 CONTROL: Southwest 1248, turn further left heading of three . . . er, correction. . . a two niner zero intercept three one centre localizer.

19:02:35.1 FIRST OFFICER: Heading two nine zero to join three one centre loc Southwest 1248.

19:02:39.0 CAPTAIN: Two nine zero to intercept.

19:02:40.2 FIRST OFFICER: Two ninety.

19:04:32.5 CAPTAIN: All right, here we go.

19:04:32.6 APPROACH CONTROL: Southwest 1248, braking action [on runways] reported fair except at the end [where] it's poor.

19:04:37.1 FIRST OFFICER TO APPROACH CONTROL: Okay, thanks.

19:04:38.8 CAPTAIN: We got fair in there, right?

19:04:40.2 FIRST OFFICER: Yeah. I think you look good and clean, my man.

19:04:58.0 CAPTAIN: All right.

19:06:44.2 APPROACH CONTROL: Southwest 9052, Chicago approach. Intercept three one centre localizer. Victor's current the runway three one centre RVR now four thousand five hundred.

19:06:51.5 FIRST OFFICER: Forty-five hundred.

19:07:04.1 CAPTAIN: All right. Slowin'.

19:07:05.1 APPROACH: Southwest 9052, last report I had for runway three one centre on the braking was braking fair except at the end [where] it was poor.

19:07:10.5 CAPTAIN: Go flaps to five, please.

19:07:12.9 FIRST OFFICER: Flap-o de cinco.

19:07:19.1 APPROACH: Southwest 9052, you copy last?

19:07:20.8 SW9052 TO APPROACH: I'm sorry, sir, no.

19:07:22.3 APPROACH: Yeah, braking action on runway three one centre is fair and then poor at the end . . . and we're just havin' a [Cessna] Citation land now I'll get a new pilot report [on the runway conditions].

19:08:45.0 CAPTAIN: All right landing gear down here, sir.

19:08:47.8 FIRST OFFICER: Landing gear down.

19:08:50.0 CAPTAIN: Down to twenty-five [2,500] here we go.

19:08:58.5 CAPTAIN: Ah, we're close.

19:09:41.6 CAPTAIN: All right let's go flaps to, ah, fifteen.

19:09:43.8 FIRST OFFICER: Flaps fifteen.

19:09:57.6 TOWER: Southwest 1248, Midway tower. Continue for [runway] three one centre. The winds [are] zero nine zero at nine. Brakin' action reported good for the first half, poor for the second half.

19:10:06.2 FIRST OFFICER: Thank you.

19:10:25.0 CAPTAIN: Let's go flaps to thirty, sir.

19:10:27.0 FIRST OFFICER: Flaps thirty.

19:10:38.3 CAPTAIN: Good old tailwind . . .

19:10:39.9 FIRST OFFICER: Yeah.

CAPTAIN: All right flaps forty.

19:10:43.5 FIRST OFFICER: Flaps forty.

19:10:56.1 FIRST OFFICER: Speedbrake.

19:10:57.1 CAPTAIN: Armed green light.

19:10:58.1 FIRST OFFICER: Landing gear.

19:10:59.2 CAPTAIN: Down three green.

19:10:59.9 FIRST OFFICER: Flaps.

19:11:00.4 CAPTAIN: Forty green light.

19:11:01.4 FIRST OFFICER: Before landing checks complete.

19:11:03.9 CAPTAIN: Thank you.

19:11:04.4 FIRST OFFICER: No landing clearance yet.

19:11:05.9 CAPTAIN: Nope.

19:11:51.9 CAPTAIN: All right yeah, you're right – thousand feet. One thirty-two, sink is, ah, eight fifty.

19:12:00.3 FIRST OFFICER: We're all counting on you.

19:12:01.6 CAPTAIN: Uhmhmm.

19:12:16.9 CAPTAIN: Never autobraked here, huh?

19:12:18.3 FIRST OFFICER: Yeah. Hang on tight [sound similar to laughter].

19:12:21.6 CAPTAIN: Yeah.

19:12:25.3 FIRST OFFICER: Five hundred [feet].

19:12:26.6 FIRST OFFICER TO TOWER: Landing clearance for Southwest 1248.

250

19:12:28.4 TOWER: Southwest 1248, runway three one centre cleared to land. Wind zero nine zero at nine, brakin' action fair to poor.

19:12:35.3 FIRST OFFICER: Four hundred.

19:12:36.3 CAPTAIN: All right.

19:12:37.2 FIRST OFFICER: Five green lights cleared to land.

19:12:41.0 FIRST OFFICER: Approaching minimums.

19:12:42.4 [Sound of thump].

19:12:56.5 FIRST OFFICER: One hundred [feet]. Fifty. Thirty. Ten.

19:13:07.8 [Sounds similar to aircraft touchdown]

19:13:08.5 CAPTAIN: Oh, baby, I guess it comes on.

19:13:11.5 CAPTAIN: Come on, baby.

19:13:13.4 FIRST OFFICER: About two thousand feet [of runway] to go.

19:13:14.7 CAPTAIN: Feel it.

19:13:15.9 FIRST OFFICER: You jumpin' on the . . . ?

19:13:16.3 CAPTAIN: Son of a bitch.

19:13:17.3 FIRST OFFICER: Jump on the brakes, are ya?

19:13:18.4 CAPTAIN: Ah huh.

19:13:19.5 FIRST OFFICER: I'm ahnna.

19:13:21.2 FIRST OFFICER: Whaddya.

19:13:22.3 CAPTAIN: Shit.

19:13:23.2 [Sound similar to double clunk].

19:13:23.4 CAPTAIN: Get that back there.

19:13:25.3 CAPTAIN: We ain't goin' [make it], man.

19:13:27.5 FIRST OFFICER: We're fucked.

19:13:28.7 CAPTAIN: We are fucked.

19:13:30.6 CAPTAIN: All right, keep it straight.

19:13:31.2 [Sound similar to increased engine noise]

19:13:35.0 FIRST OFFICER: Shit.

19:13:35.4 CAPTAIN: Hang on.

19:13:35.9 FIRST OFFICER: Hang on.

19:13:36.5 [Sound similar to impact].

19:13:39.0 [Sound similar to impact].

19:13:39.4 FIRST OFFICER: Oh, shit.

19:13:42.2 FIRST OFFICER: [Sound similar to groan]

19:13:43.3 CAPTAIN: [Sound similar to grunt] Fuck me.

19:13:46.5 FIRST OFFICER: Fuck.

19:13:47.1 CAPTAIN: Fuck.

19:13:48.1 [Sound similar to stick shaker].

19:13:49.1 [Sound similar to chime].

19:13:51.0 [Sound similar to clunk].

19:13:51.4 FIRST OFFICER: Southwest 1248 went over the end.

TOWER: Say again.

19:13:54.3 FIRST OFFICER: We went off the end of the runway.

19:13:54.4 CAPTAIN: Shut down.

19:13:58.4 FIRST OFFICER: Fuck.

19:14:00.6 CAPTAIN: Shuttin' down.

END OF RECORDING

The aircraft had rolled through a blast fence, an airport perimeter fence and onto an adjacent roadway, where it struck an automobile before coming to a stop. A child in the automobile was killed, one automobile occupant received serious injuries and three other automobile occupants received minor injuries. Eighteen of the 103 aeroplane occupants (98 passengers, 3 flight attendants and 2 pilots) received minor injuries, and the aeroplane was substantially damaged.

PHILADELPHIA, Pennsylvania, USA

7 February 2006

At the start of the second day of a five-day, eight-leg trip sequence for its flight crew, United Parcel Service company Flight 1307, a McDonnell Douglas DC-8–71F, registered as N748UP, pushed back from the gate at Hartsfield-Jackson Atlanta International Airport, Atlanta, Georgia, en route to Philadelphia. The time was about 10.41 p.m. The first officer was flying the aeroplane and everything went well until just before the descent to Philadelphia. At 11.34 p.m. the aeroplane was descending through 31,000 feet, about 50 nautical miles south-west of Washington, DC, when the first officer smelled an odour 'like wood burning'.

23:34:39 FIRST OFFICER: Smells like wood burning. Smell that?
23:34:42: FLIGHT ENGINEER: Hell, yeah. I smelled it for a couple of seconds.
23:35:40: FIRST OFFICER: It's pretty strong now.
23:35:54 FLIGHT ENGINEER: It's more in the back.
23:36:23 CONTROL: UPS 1307 contact Washington Centre [on frequency] one two five point four five.

23:36:28 CAPTAIN: Two five four five UPS 1307, thank you.

23:36:31 CONTROL: Good day.

23:36:39 CAPTAIN: You might try turning like a [battery] pack off, uh, [and] see if that makes any difference.

23:37:18 CAPTAIN TO CONTROL: Uh, centre good day. It's, uh, UPS 1307 with you passing 27,500 for 24,000.

23:37:23 CONTROL: UPS 1307, Washington Centre, roger.

23:38:58 [Sound of tone similar to altitude alert]

23:39:00 FIRST OFFICER: Twenty-five [to] twenty-four [thousand feet].

23:39:01 CAPTAIN: Twenty-five twenty-four.

23:39:16 FIRST OFFICER: Why would it . . . smell like wood?

23:39:19 FLIGHT ENGINEER: Yeah, it does smell like wood. It doesn't smell electrical smell . . .

23:39:26 FIRST OFFICER: That's just duct . . . that's just duct work but there's no wood.

23:39:30 CAPTAIN: Yeah.

23:39:30 FIRST OFFICER: There's no brace around any of that stuff, is there?

23:39:34 CAPTAIN: No.

23:39:36 FIRST OFFICER: . . . It's like a wood-burning kit.

23:41:04 CONTROL: UPS 1307, cross ten miles south of Woodstown at and maintain one one thousand [feet]. Philly altimeter three zero zero five.

23:41:11 CAPTAIN TO CONTROL: Ten south of Woodstown at one one thousand, UPS 1307.

23:43:18 CAPTAIN: [You] might try those bleeds, uh, switches too, . . .

23:43:22 FLIGHT ENGINEER: . . . fume evacuation.

23:43:26 CAPTAIN: Yeah.

23:43:27 FLIGHT ENGINEER [reading manual]: . . . said put the pack to max flow.

23:43:28 CAPTAIN: Okay.

23:43:29 FLIGHT ENGINEER: Set pack to max flow. Recirc[ulation] fan off? Recirc[ulation] fan off.

23:43:34 CAPTAIN: [The smell] may be coming from one of those bleeds, you know.

23:43:46 FIRST OFFICER: I got go-around set in there. Visual backed up with the ILS two seven right one oh eight-ninety five eleven touchdown zone . . . Three zero zero five approach check.

23:44:14 FLIGHT ENGINEER: Approach checklist. Pressurization set. [Thrust] Reverser shut-off switch?

23:44:22 CAPTAIN: Open.

23:44:38 CONTROL: UPS 1307 contact Philly approach one two four point three five.

23:44:39 CAPTAIN TO CONTROL: Two four three five UPS 1307, good day.

23:44:42 CONTROL: Good day.

23:44:48 FLIGHT ENGINEER: Approach brief?

23:44:49 FIRST OFFICER: Complete.

23:44:50 FLIGHT ENGINEER: Altimeter?

23:44:52 CAPTAIN: Thirty oh five set.

23:44:55 FLIGHT ENGINEER: Thirty oh five set approach checklist complete.

23:44:59 CAPTAIN TO APPROACH: And Philly approach good day. It's UPS 1307 heavy with you passing 15,000 [feet] for 11,000.

23:45:19 APPROACH: UPS 1307 heavy Philly runway 27 right, altimeter three zero zero four.

23:45:25 CAPTAIN TO APPROACH: Thirty oh four two seven right UPS 1307 heavy.

23:46:29 APPROACH: UPS 1307 heavy fly heading of zero five zero and descend and maintain six thousand.

23:46:33 CAPTAIN TO APPROACH: Zero five zero down six thousand UPS 1307 heavy.

23:47:02 CAPTAIN: Can you still smell it in the back there?

23:47:06 FLIGHT ENGINEER: Yeah, it smells like it was more to the back there.

2347:14 CAPTAIN: Smell like it more (strong) back there.

23:47:18 FIRST OFFICER: Smells like cardboard burning, doesn't it? You didn't see smoke, though, something like that?

23:47:29 [Sound similar to cockpit door operating]

23:47:58 [Sound similar to cockpit door operating]

23:47:59 FLIGHT ENGINEER: It is definitely stronger in the back.

23:48:01 CAPTAIN: Is that right?

23:48:02 FLIGHT ENGINEER: Yeah, it is definitely stronger in the back.

23:48:04 CAPTAIN: Huh.

23:48:07 FLIGHT ENGINEER: Well, [there] does not appear to be any smoke or haze.

23:48:10 CAPTAIN: What's that [you said]?

23:48:10 FLIGHT ENGINEER: I just shined my light back there [and] I can't see any haze or anything.

23:48:21 CAPTAIN: Did you, did you try all the bleeds?

23:48:23 FLIGHT ENGINEER: Well, I tried the second one off now maybe another one off.

23:48:27 CAPTAIN: Okay.

23:48:27 APPROACH: UPS 1307 heavy fly heading zero six zero descend and maintain four thousand.

23:48:31 CAPTAIN TO APPROACH: Zero six zero down to four thousand UPS 1307 heavy.

23:48:41 FIRST OFFICER: Four thousand . . .

23:52:30 CAPTAIN: Does [the smoke] seem to get any better? Uh, checklist?

23:52:35 FIRST OFFICER: Flaps ten please.

23:52:41 FLIGHT ENGINEER: Pack smoke warning.

23:52:42 CAPTAIN: Yeah, that's what I am doing under the pack smoke . . . Fumes evacuation.

23:52:47 FLIGHT ENGINEER: Smoke, yeah? No smoke detectors.

23:52:52 CAPTAIN: Yeah, yeah no smoke detectors going off.

23:53:19 FIRST OFFICER: Flaps twenty-five.

23:54:07 APPROACH: UPS 1307 heavy, turn left to zero one zero [and] descend and maintain two thousand one hundred [feet].

23:54:12 CAPTAIN TO APPROACH: Zero one zero down to two point one UPS 1307 heavy.

23:54:20 FIRST OFFICER: Two thousand one hundred.

23:54:42 FLIGHT ENGINEER: Okay, we got cargo smoke.

23:54:45 CAPTAIN: You got cargo smoke? Let's do that checklist if you got time.

23.54:52 FIRST OFFICER: All right, I am turning into the airport then.

CAPTAIN TO APPROACH: UPS 1307 heavy has the field in sight.

23:55:01 APPROACH: UPS 1307 heavy cleared for the visual approach runway 27 right contact the tower one eighteen five.

23:55:06 CAPTAIN TO APPROACH: Eighteen five. See ya.

23:55:11 CAPTAIN TO TOWER: Tower good day. It is UPS 1307 heavy with you visual for 27 right.

23:55:15 TOWER: 1307 heavy 27 right wind two six zero at six cleared to land.

23:55:19 CAPTAIN TO TOWER: Cleared to land and, uh, listen we just got a cargo smoke indicator come on. Can we have the [fire fighting] equipment meet us?

23:55:27 TOWER: Okay. I'll do that for you. Cargo smoke indicator. [Sound of alarm in background] Uh, uh, just [how many] souls on board, amount of fuel, sir?

23:55:34 CAPTAIN TO TOWER: Uh, three souls [and] two hours of fuel.

23:55:37 TOWER: Two hours of fuel roger, [is] that, sir? We are bringing [the fire fighting equipment] out now.

23:55:40 CAPTAIN TO TOWER: Thanks.

23:55:45 FIRST OFFICER: Gear down.

23:55:46 CAPTAIN: Gear.

23:55:48 [Sound similar to landing gear operation]

23:55:48 FLIGHT ENGINEER: Okay, it's showing that we have a lower aft cargo fire [in cargo] section C.

23:55:58 CAPTAIN: Oxygen masks on if you don't have 'em and run through that checklist . . . by yourself, okay?

23:56:03 FLIGHT ENGINEER: Okay, [we have a] lower and, uh, or main cargo fire. Oxygen masks on a hundred per cent.

23:56:12 TOWER: UPS 1307 heavy is cleared to land runway 27 left. The wind is two six zero at six.

23:56:18 FIRST OFFICER TO TOWER: Cleared to land 27 left UPS 1307.

23:56:22 FIRST OFFICER TO TOWER: Okay, I got . . .

23:56:22 TOWER: I am sorry – last call on ground, say again.

23:56:24 FIRST OFFICER: I got packs one off one on min flow. Recirc fan is off. Oxygen air diffuser valve open . . .

23:56:35 TOWER: When you get a chance I know you are busy 1307 heavy [but] can you give me fuel in pounds, please?

23:56:44 CAPTAIN TO TOWER: 21,700 [pounds]

23:56:46 TOWER: Thank you.

2356:48 FIRST OFFICER: Flaps thirty five.

23:56:50 FLIGHT ENGINEER: Okay, main cargo air shutoff valve is closed I gotta go in the back and do that.

23:56:56 [Sound similar to cockpit door operating]

2356:57 FIRST OFFICER: Smelling pretty good now.

23:56:58 CAPTAIN: Yeah.

23:57:02 FIRST OFFICER: Flaps full.

23:57:04 [Sound similar to cockpit door operating]

23:57:09 FLIGHT ENGINEER: Yeah, we're going to have to do an evacuation, okay? Tell them we are going to have to do an evacuation when we get down. I got that valve shut off back there. There is smoke. [The] radio pack blower switch [is] off.

23:57:19 FIRST OFFICER: We are almost on the ground.

23:57:44 FLIGHT OFFICER: Land as soon as possible.

23:57:47 FIRST OFFICER: Landing checklist when you get a chance. Cleared to land?

23:57:59 CAPTAIN: Five hundred feet [and] speed sink is eight.

23:58:03 [GPWS voice] *Glideslope.*

23:58:04 [GPWS voice] *Five hundred.*

23:58:06 FLIGHT ENGINEER: Ignition. Gear. Anti-skid.

23:58:08 TOWER: 1307 heavy, just confirmed you are lined up for the left side. It appears you are lined up for the right.

23:58:13 CAPTAIN TO TOWER: I'm sorry. I thought we were cleared for the right. Uh, are we cleared to land on the right?

23:58:16 TOWER: You are cleared to land on the right. We will just tell fire [of the change].

23:58:20 CAPTAIN TO TOWER: Okay.

23:58:23 [GPWS voice] *Glideslope Glideslope Glideslope.*

23:58:50 [GPWS voice] *One hundred.*

23:58:54 CAM. [GPWS voice] *Fifty.*

23:58:55 [GPWS voice] *Thirty.*

23:58:56 [GPWS voice] *Twenty.*

23:58:57 [GPWS voice] *Ten.*

23:59:00 [Sound similar to spoiler handle movement]

23:59:00 TOWER: He is rolling out, flaring as we speak.

23:59:0 FLIGHT ENGINEER: Okay, we got smoke in the cockpit now.

23:59:05 [Sound similar to thrust reverser deployment and engine acceleration]

23:59:11 CAPTAIN: Eighty [knots].

23:59:13 FLIGHT ENGINEER: Tell 'em we have smoke in the cockpit. We are evacuating.

23:59:14 CAPTAIN: Seventy [knots].

23:59:18 FIRST OFFICER: You have the aircraft.

23:59:21 [Sound similar to window(s) being operated]

23:59:25 FIRST OFFICER: Okay, emergency evacuation.

23:59:26 CAPTAIN: Evacuation checklist.

23:59:29 [Sound similar to coughing]

23:59:30 CAPTAIN: Parking brake set.

23:59:33 FIRST OFFICER: Fuel shut off levers off.

23:59:35 CAPTAIN: Battery switch battery.

23:59:38 FIRST OFFICER: UPS thirteen oh seven evacuating the aircraft.

23:59:41 TOWER: Two seven right is closed and the air crew is evacuating the aircraft.

23:59:41 FIRST OFFICER: Fire handles full forward.

23:59:45 END OF RECORDING

The aeroplane landed on runway 27R at about 11.59 p.m. The first officer called for an emergency evacuation. All of the flight crew members successfully evacuated the aeroplane using the emergency slide located at the left forward door.

LEXINGTON, Kentucky, USA
27 August 2006

On 27 August 2006, at about 6.06 a.m., Comair Flight 5191, a Bombardier CL-600–2B19, registered as N431CA, took off from Blue Grass Airport in Lexington, Kentucky, bound for Atlanta, Georgia. The flight crew had checked in for the flight at 5.15 a.m. They were casually conversing as they collected their flight release paperwork, which included weather information, routine safety-of-flight notices to airmen, the tail number of the aeroplane to be flown and the flight plan to their destination. They then proceeded to an area on the air carrier ramp where two Comair Canadair regional jet aeroplanes were parked. A Comair ramp agent, who was performing a security check of the aeroplane, noticed that the flight crew had boarded the wrong aeroplane and had started its auxiliary power unit. Another company ramp agent notified the crew that they had boarded the wrong aeroplane. They then shut down the power unit and walked to the correct aeroplane. The airport air traffic control tower that morning was staffed by one controller, who was responsible for all tower and radar positions.

The first officer was the flying pilot.

05:38:04.3 CAPTAIN: Circuit breakers, checked, nose wheel steering is off, hydraulic pumps, all off, landing gear lever, down, spoiler lever, lever, retracted, flaps lever, set to flaps twenty, radar, off, AHRS, mag, landing gear manual release, ADG manual release, battery master, on, fire protection, checked, nav lights, on, external . . . A/C, no APU, hydraulic pump 3A is on, nose-wheel door, closed, aircraft/ew docs, on board, flight compartment safety inspection is complete.

05:38:40.9 [Three tones similar to CVR test tone]

05:38:53.0 [Sound of person whistling]

05:38:55.7 [Sound of hi-lo chime]

05:39:07.6 [Sound of chimes similar to fire protection fire test signal] smoke.

05:39:24.9 CAPTAIN: I was talking to another guy I flew with yesterday. He was, uh, had put his, bid in for, uh, JFK captain. He wasn't real happy about it but . . .

05:39:35.4 FIRST OFFICER: First officer?

05:39:36.3 CAPTAIN: Yeah.

05:39:37.9 FLIGHT ATTENDANT: Would you turn the smoking–seat belt sign on for me?

05:39:40.2 [Sound of chime]

05:39:41.0 CAPTAIN: You got it. Here come the lights.

05:39:43.5 FIRST OFFICER: Why wasn't he too happy about it? He can always change it.

05:39:46.8 CAPTAIN: Yeah, you know he just, he's just not really looking forward to reserves, that's all. But he feels like, you know, and I think he's right if he wants to get out of here. That's his decision he wants to do but he's gotta get that PIC [pilot-in-command?] . . .

05:40:01.1 FIRST OFFICER: Exactly.

05:40:01.9 CAPTAIN: . . . You gotta bite the bullet sometimes but, I mean, nobody wants to do reserve there.

05:40:08.5 FIRST OFFICER: Nope, not here. Not the way they do it . . .

05:40:21.9 FIRST OFFICER: You know, you're on for six days and you might fly eight hours 'cause they . . .

05:40:27.1 CAPTAIN: It's amazing though right now, they are using everybody pretty efficiently. Just shows you what they can do. Like, I mean, I don't have more than ten hours in a hotel, any of these days that I've been on . . .

05:40:38.2 FIRST OFFICER: Really.

05:40:38.7 CAPTAIN: . . . and it's been that way for all month. Now September rolls around and I'll guarantee you it'll be a different story.

05:40:50.3 FIRST OFFICER: . . . because I know Cincinnati base, they have a lot of reserves. But I understand.

05:40:56.0 CAPTAIN: Then, they send them all to New York.

05:40:57.1 FIRST OFFICER: Right, exactly.

05:40:58.4 CAPTAIN: Yeah.

05:40:59.9 [Sound similar to stick shaker test]

05:41:04.6 [Sound of test *Glideslope Whoop, Whoop, Pull up, Wind shear, Wind shear, Wind shear, Terrain, Terrain, Whoop, Whoop, Pull up*]

05:41:09.4 [Sound of person whistling]

05:42:10.8 CAPTAIN: Funny you were talking about . . . I mean, I, I, I flew with a guy who was, he said he filled out the application process, he filled out the application and went through the background checks . . . He was telling me all about it.

05:42:33.3 FIRST OFFICER: Well, they might do all that stuff prior to actually, giving you, the interview date.

05:42:38.5 CAPTAIN: Uh huh.

05:42:39.3 FIRST OFFICER: But you are by no means guaranteed anything . . .

05:42:45.1 CAPTAIN: I just talked to my wife about it. We looked at. We looked at it on line, you know, and I was looking at the pay scales and, uh, yeah, I know they provide a place for you to live and things like that and was at four thousand, I don't know, forty-four hundred dollars.

05:43:01.8 FIRST OFFICER: Fifty-two, fifty-two twelve, a month for the first . . . month tax free.

05:43:09.4 CAPTAIN: Yeah, the last time I looked at it or it was like forty-five or something and . . . but . . . I talked to a guy who was in the military, [and] he said he . . . said it's really pretty for a desert, you know it's . . .

05:43:22.8 FIRST OFFICER: Yeah, well, there's a guy, a military guy, up, uh, a first officer, in Kennedy, [and] he's like I think you're doing the right thing. He says if not to visit, maybe to be an expatriate and live there is not a good thing.

05:43:40.9 CAPTAIN: Yeah, what I heard, you know you can't buy land. They'll let you buy a condo, like on a high rise or something, thank you, but you can't buy property . . .

05:44:04.9 FIRST OFFICER: Yeah, howdy, like yeah, yeah . . . and then I kept thinking about it . . . I guess, when I'm, I'm deciding on making a major decision, if it doesn't feel right in my gut. Or if I don't have a little voice, if it starts talking to me and I'm like I need to re-evaluate.

05:44:29.5 CAPTAIN: Yeah.

05:44:43.2 FIRST OFFICER: You know, it'd be nice to go over there and fly heavy metal, fly international, but they work you hard over there I've been told.

05:44:50.3 CAPTAIN: Oh do they?

05:44:51.1 FIRST OFFICER: Yeah, they fly you if they can up to a hundred hours . . . They have triple sevens . . . Like for you with the kids, you'd get a housing allowance at a villa. And for me and my wife with no kids, we'd get an apartment. The apartments don't allow any animals and I have four dogs and I'm not, I'm not about to give up. I've had 'em for a while. If I fly overseas, I wanna start and finish here in the States.

05:45:36.1 CAPTAIN: You were overseas already. Is that what you said?

05:45:36.9 FIRST OFFICER: No, if I, if I did fly overseas.

05:45:38.8 CAPTAIN: Ah, okay.

05:45:44.8 CAPTAIN: Emergency equipment, checked, crew oxygen masks, checked left and right . . . CVR, checked, standby instruments, checked, fire protection, checked, gravity cross-flow, checked, duct monitor, checked, hydraulics, auto and on, ice detector, has been checked, cabin signs are on, emergency lights are armed, stall protection system, checked, anti-skid, checked and armed, MLG bay overheat, checked, stab/Mach trim, engaged, engine controls, checked, aileron rudder trims

268

checked yaw damper engaged, cabin/exterior checks complete gear and safety pins, have been removed. Acceptance checklist is complete.

05:46:19.5 FIRST OFFICER: I started looking at it a little more. There was just too many . . . to get through.

05:46:31.4 FIRST OFFICER: Is this first officer single?

05:46:34.8 CAPTAIN: I don't think so but his, his name is, he's got an Arab, uh . . .

05:46:40.5 FIRST OFFICER: Oh.

05:46:42.3 CAPTAIN: He got, ah, he has some kind of Arab name . . . or something, ah.

05:46:50.9 CAPTAIN: He might blend in a little bit but I heard it's like sixty or seventy per cent European, I mean.

05:46:55.2 FIRST OFFICER: Well, it's not even owned by United Arab Emirates. It's owned by a British company.

05:47:01.1 CAPTAIN: Oh really.

05:47:08.9 FIRST OFFICER: Yeah, but you gotta deal with a lot of Brits and Australians. You know some of these Brits are a little uptight.

05:47:34.5 FIRST OFFICER: If circumstances were different I, I'd consider it.

05:47:55.9 FIRST OFFICER: I'm gonna talk to my dad to see if maybe he can help me out but I think I'm gonna invest in a seven three type rating. If I make captain here. I need like three hundred PIC [hours] to be eligible to meet their requirements and then I'll . . .

05:48:14.1 CAPTAIN: Who's that, Southwest? Isn't that three thousand PIC?

05:48:16.7 FIRST OFFICER: Fifteen hundred.

05:48:18.6 CAPTAIN: Fifteen hundred?

05:48:18.7 FIRST OFFICER: Actually no, thirteen hundred PIC.

05:48:20.8 CAPTAIN: Oh, okay.

05:48:21.1 FIRST OFFICER: Like thirteen hundred, that's a weird number.

05:48:24.4 ATIS: Lexington Bluegrass information Alpha, 0854 automated weather. Wind one niner zero at eight, visibility eight, few clouds six thousand, sky broken niner thousand. Temperature two four, dew point one niner, altimeter three zero zero zero. ILS and visual approaches in use. Landing and departing runway 22. Runway 22 glideslope out of service. Pilots use caution for construction on air carrier ramp. Hazardous weather information available on HIWAS, Flight Watch or Flight Service frequencies. All departures contact ground control on one two one point niner. Advise you have information Alpha.

05:49:42.2 FIRST OFFICER: Clearance good morning. Comair 191's going to Atlanta with Alpha.

05:49:49.3 CONTROL: Comair 191, Lexington clearance. Cleared to Atlanta Airport via Bowling Green, ERLIN TWO arrival. Maintain six thousand. Expect flight level two seven zero one zero minutes after departure. Departure's one two zero point seven five. Squawk six six four one.

05:50:06.5 FIRST OFFICER: Okay, got uh, Bowling Green, uh, missed the other part. Six thousand, twenty point seven five. Six six four one.

05:50:14.1 CONTROL: Comair 191, it's ERLIN TWO, Echo Romeo Lima, India, November Two arrival.

270

05:50:20.4 FIRST OFFICER: 'Kay ERLIN Two, 'preciate it, Comair 191.

05:50:53.7 CAPTAIN: Direct Bowling Green, Bowling Green the ERLIN TWO. Is that good?

05:50:58.4 FIRST OFFICER: Any easier than that.

05:50:59.9 [Sound of laughter]

05:51:29.3 CAPTAIN: Chattanooga looks good for the alternate.

05:51:29.8 DEADHEADING PILOT RIDING IN THE CABIN [term used when pilot is transported by another plane to a location where s/he will fly a future flight]: Well, how's it going, guys?

05:51:30.5 FIRST OFFICER: Dude, what's up?

05:51:32.6 DEADHEAD: How you doin'?

05:51:33.3 CAPTAIN: Hey, good, how you doin'?

05:51:34.8 DEADHEAD: I'm . . . with Air Tran trying to get a lift to work this morning.

05:51:37.1 CAPTAIN: Hey, no problem. [Take] any seat . . . you already got one.

05:51:38.8 DEADHEAD: There you go.

05:51:40.0 CAPTAIN: Beauty.

05:51:43.2 DEADHEAD: All right, sir.

05:51:44.4 CAPTAIN: 'Preciate it.

05:51:45.1 DEADHEAD: You bet.

05:51:46.1 CAPTAIN: Any time.

05:51:47.2 DEADHEAD: Thanks a lot.

05:53:00.2 FLIGHT ATTENDANT: Passenger's request for an electric cart in the gatehouse for a passenger.

05:53:03.7 CAPTAIN: Can I grab another Coke from you?

05:53:05.2 FLIGHT ATTENDANT: Would you like one?

05:53:06.1 FIRST OFFICER: No thanks.

05:55:04.4 FIRST OFFICER: . . . My wife called a little excited about it but . . . yeah, you'd be gone like three or four days at a time, which you know with your wife and kids it might be a little difficult.

PUBLIC ADDRESS: Ladies and gentlemen from the flight deck, like to take this time to welcome you also on board Comair Flight 5191 direct flight to Atlanta. We'll be cruising at, uh, twenty-seven thousand feet this morning. And once we do get in the air, it looks like one hour and seven minutes en route. Distance of travel today, we got we got four hundred and twenty-two miles. Weather conditions Atlanta . . . some light winds out of the east. Looks like some broken clouds and current temperature of seventy-two degrees Fahrenheit. We'll try to keep it as quiet as possible. Hopefully you can catch a nap going into Atlanta. It's our pleasure having you all on board.

05:56:14.0 CAPTAIN TO FIRST OFFICER: For our own briefing, Comair standard. Run the checklist [at] your leisure. Keep me out of trouble. I'll do the same for you. I don't jump on the brakes on your landing. I'll follow along with you. Just let me know when you want me to take it. That's it.

05:56:23.8 [Sound of two clicks similar to pilot seat adjustment]

05:56:25.2 FIRST OFFICER: I'll do the same whenever you're ready.

05:56:26.9 CAPTAIN: All right.

05:56:27.6 GROUND CONTROL: Information Bravo is now current. The altimeter's three zero zero zero.

05:56:30.4 CAPTAIN: Sounds good.

05:56:34.1 FIRST OFFICER: Right seat flex takeoff procedures off of, um . . . he said what runway? One of 'em . . . 24.

05:56:43.4 CAPTAIN: It's 22.

05:56:45.9 FIRST OFFICER: One ninety at eight.

05:56:49.9 FIRST OFFICER: Two two up to six, white data . . . FMS, flaps twenty . . . smokes or breaks come back here. Come into four or two two. On two two the ILS is out. Or the glideslope is, the REILS are out. The, uh, came in the other night it was like [sound similar to audible exhale] lights are out all over the place.

05:57:07.8 CAPTAIN: All right.

05:57:08.4 FIRST OFFICER: Right. Remember this runway [is] predicated, before we just go back to Cincinnati.

05:57:12.9 CAPTAIN: Okay.

05:57:13.7 FIRST OFFICER: Umm, no continuous, anti-ice, weather radar, hand fly 'til about ten. Taxi instructions with ATC.

05:57:21.7 CAPTAIN: All right.

05:57:23.3 FIRST OFFICER: Let's take it out and, um, take uuuh, [taxiway] Alpha. [Runway] 22's a short taxi.

05:57:31.1 CAPTAIN: Yeah.

05:57:35.4 FIRST OFFICER: Any questions?

05:57:36.5 CAPTAIN: No questions. Before starting at your leisure.

05:57:38.4 FIRST OFFICER: ACM crew briefing.

05:57:39.3 CAPTAIN: Complete.

05:57:40.0 FIRST OFFICER: Takeoff brief.

05:57:40.4 CAPTAIN: Complete.

05:57:40.7 FIRST OFFICER: Radios, NAVaids.

05:57:42.0 CAPTAIN: Uh, six thousand, your side, both in white data, confirmed the flight plan. We got, uh, tower, ground twenty one . . . Is everything on one? Do you know?

05:57:49.7 FIRST OFFICER: No, it's not.

05:57:53.5 CAPTAIN TO FLIGHT ATTENDANT: If we do have an emergency, I like to open the door and talk face to face.

05:58:04.5 FLIGHT ATTENDANT: Okay.

05:58:06.0 CAPTAIN: That's about it. Anything for me?

05:58:07.0 FLIGHT ATTENDANT: Sounds great.

05:58:07.9 CAPTAIN: All right.

05:58:08.4 FLIGHT ATTENDANT: See you in Atlanta.

05:58:12.2 CAPTAIN: Uuh, start engines your leisure.

05:58:17.1 FIRST OFFICER: Takeoff brief.

05:58:18.3 CAPTAIN: Hey man, we already did that one.

05:58:20.5 FLIGHT ATTENDANT: We did?

05:58:21.0 CAPTAIN: Yeah.

05:58:21.2 FIRST OFFICER: I'm sorry.

05:58:22.2 [Sound of laughter]

05:58:23.3 FIRST OFFICER: I'll get it to ya.

05:58:24.1 CAPTAIN: [Sound of laughter]

05:58:24.6 FIRST OFFICER: Papers, manifest.

05:58:26.0 CAPTAIN: It's complete out the door.

05:58:27.2 FIRST OFFICER: Fuel quantity.

05:58:27.4 CAPTAIN: Required to have [seventy], seventy-three.

05:58:33.3 [Sound of several clicks similar to cockpit door operation]

05:58:35.1 FIRST OFFICER: V speeds, takeoff data fifty temp. V one, V R is thirty-seven forty-two . . . forty-five [spoken at a very fast speed] you got normal thrust . . . point two.

05:58:43.5 CAPTAIN: Set for flaps twenty.

05:58:44.5 FIRST OFFICER: Doors.

05:58:44.9 CAPTAIN: Closed.

05:58:45.3 FIRST OFFICER: Beacon.

05:58:45.6 CAPTAIN: On.

05:58:46.0 FIRST OFFICER: Fuel pumps.

05:58:46.8 CAPTAIN: Number one's on.

05:58:48.7 FIRST OFFICER: See this is telling me that I really need to do something.

05:59:05.7 FIRST OFFICER TO CONTROL: And ground, Comair 191, just a heads-up on the push [back].

05:59:11.1 GROUND: Comair 191 advise ready to taxi.

05:59:13.3 CONTROL: Roger.

05:59:47.2 FIRST OFFICER: How long of a ride is that?

05:59:50.9 CAPTAIN: Uuuuh, maybe, maybe an hour.

05:59:57.5 FIRST OFFICER: That's cool.

06:00:09.4 CAPTAIN: . . . both kids were sick, though, they, well they all got colds. It was an interesting . . . dinner last night.

06:00:16.1 FIRST OFFICER: Really.

06:00:16.6 CAPTAIN: Huh, oh gosh.

06:00:19.1 FIRST OFFICER: How old are they?

06:00:20.0 CAPTAIN: Three months and two years old

. . . Who was sneezing, either nose wiped, diaper change. I mean, that's all we did all night long.

06:00:31.0 FIRST OFFICER: Oh yeah, I'm sure.

06:00:34.9 FIRST OFFICER: That's a nice range, age range.

06:00:37.7 CAPTAIN: Yeah, I like two years apart basically and that's kinda what we were going for.

06:00:45.3 CAPTAIN: My wife wants four, I, I, I'm, I was good at one.

06:00:48.5 FIRST OFFICER: She wants four?

06:00:50.1 CAPTAIN: Yeah.

06:00:50.2 [Sound of chuckle]

06:00:52.7 FIRST OFFICER: It'd be like, honey . . .

06:01:02.7 [Sound of chime]

06:01:07.1 FIRST OFFICER: Yeah, it's especially being on reserve it, it's gotta be tough being away.

06:01:12.2 CAPTAIN: Ah, tough on her, oh my God. That's why she came down yesterday. She's, like, I just need to get out of this house.

06:01:18.0 FIRST OFFICER: Yeah, I bet.

06:01:18.9 CAPTAIN: I'm, like, I understand. I, I told her, why don't you just spend the night. She said, Well, if you're gonna get up at oh-dark-thirty and she said you'll end up waking up the babies. I'm like, yeah, you're probably right.

06:01:32.1 FIRST OFFICER: Yeah, it would just be like being at home.

06:01:35.5 CAPTAIN: Yeah, she's like you know, I don't know, she's like I'll . . .

06:01:38.5 FIRST OFFICER: Instead of having her rush back and drive . . .

06:01:40.6 CAPTAIN: And we got a dog.

06:01:42.9 FIRST OFFICER: Aah, trust me the dog . . . Be on the . . . slim-fast diet . . . for a night.

06:01:47.4 [Sound of laughter]

06:01:48.7 FIRST OFFICER: Uh, parking brake.

06:01:49.6 CAPTAIN: That's on.

06:01:50.0 FIRST OFFICER: Number two, actually, engine.

06:01:51.7 CAPTAIN: One and two are started.

06:01:52.5 FIRST OFFICER: Starting engines complete.

06:01:54.3 CAPTAIN: And before taxi.

06:01:55.6 FIRST OFFICER: Anti-iiiiice.

06:01:56.6 CAPTAIN: . . . probes are low and on.

06:01:58.3 FIRST OFFICER: Nose wheel steering.

06:01:59.5 CAPTAIN: That's armed.

06:02:00.0 FIRST OFFICER: Taxi check complete.

06:02:01.3 FIRST OFFICER: Comair 191 is ready to taxi. We have Alpha.

06:02:03.8 GROUND: Comair 191, taxi to runway two two. Altimeter three zero zero zero and the winds are two zero zero at eight.

06:02:08.9 FIRST OFFICER: Three triple zero and taxi [to runway 22] two two, Comair 191.

06:02:12.6 GROUND TO ANOTHER AIRCRAFT WAITING TO TAKE OFF: Eagle flight runway two two, cleared for takeoff.

06:02:15.1 CAPTAIN: Clear left.

06:02:17.3 FIRST OFFICER: [Clear] on the right.

06:02:17.9 CONTROL TO ANOTHER AIRCRAFT: Skywest 6819 radar contact, say altitude leaving.

06:02:18.9 CAPTAIN: Flaps twenty, taxi check.

06:02:21.0 FIRST OFFICER: Full right.

06:02:23.8 GROUND TO ANOTHER AIRCRAFT: Skywest 6819, climb and maintain one zero thousand, ten thousand, join Victor one seventy one . . .

06:02:24.0 FIRST OFFICER: Full left.

06:02:25.5 CAPTAIN: Test your brakes any time.

06:02:31.1 FIRST OFFICER: I want to . . . down.

06:02:32.3 CAPTAIN: Sure.

06:02:41.5 FIRST OFFICER: Let's see, comin' back.

06:02:51.6 FIRST OFFICER: Brakes.

06:02:52.3 CAPTAIN: They're checked.

06:02:53.2 FIRST OFFICER: Right, flaps.

06:02:54.4 CAPTAIN: Set twenty, indicating twenty.

06:02:55.6 FIRST OFFICER: Flight controls.

06:02:56.3 CAPTAIN: Check left.

06:02:58.3 FIRST OFFICER: On the right, trims.

06:02:59.5 CAPTAIN: Engage zero seven point two.

06:03:02.2 FIRST OFFICER: Radar terrain displays. [Spoken in a yawning voice]

06:03:04.0 FIRST OFFICER: All the taxi check's complete. [Spoken in a yawning voice]

06:03:12.0 CAPTAIN: Finish it up your leisure.

06:03:16.4 FIRST OFFICER: Yeah, I know three guys at Kennedy. Actually two guys, uh . . . and he went but he didn't get past the sim[ulator].

06:03:26.7 CAPTAIN: Oh, really?

06:03:29.1 FIRST OFFICER: And then, um, a first officer from Cinci . . .

06:03:34.5 GROUND: Eagle flight radar contact, radar contact. Say altitude leaving.

06:03:35.1 FIRST OFFICER: Got through the second part . . . What do you do the, uh, these tests . . . and he didn't, and that's as far as he got.

06:03:40.8 GROUND: Eagle flight 882, climb and maintain one zero thousand, ten thousand.

06:03:49.3 FIRST OFFICER: And then . . . he actually got offered the position.

06:03:54.5 CAPTAIN: Did he take it or . . .

06:03:55.5 FIRST OFFICER: Yeah.

06:03:56.1 CAPTAIN: Ah, okay.

06:04:01.2 FIRST OFFICER: Second engine started, anti-ice probes windshield low.

06:04:03.1 GROUND: . . . 1691, previous question.

06:04:05.6 FIRST OFFICER: Hydraulics checked, APU's on. FMS, we got runway [22] two two out of Lexington up to six [thousand feet].

06:04:13.3 FIRST OFFICER: Thrust reversers are armed, auto crossflow is manual, ignition is off, altimeters three triple zero across the board, cross-checked. I'll be in the back.

06:04:24.8 CAPTAIN: Got one.

06:04:25.7 TOWER: Skywest 6819, contact Indy Centre one two six point three seven.

06:04:29.6 SKYWEST: Two six three seven, Skywest 6819.

06:04:32.6 TOWER: Eagle flight 882, turn right heading two seven zero, join Victor one seventy-one. Resume navigation.

06:04:37.5 EAGLE FLIGHT 882: Eagle 882 two seven zero, join Victor one seventy-one . . .

06:04:38.2 FIRST OFFICER ON PUBLIC ADDRESS TO CABIN: And folks one [last] time from the flight deck, we'd like to welcome you aboard. We're going to be underway momentarily . . . Sit back, relax, enjoy the flight. Kelly [flight attendant], when you have a chance, please prepare the cabin.

06:04:49.3 FIRST OFFICER: Pre-takeoff [checklist] complete. Cabin report received CAS.

06:04:53.4 CAPTAIN: Checked and clear.

06:05:06.3 FIRST OFFICER: Cabin report's received, CAS clear . . . before takeoff check's complete, ready.

06:05:12.6 CAPTAIN: All set.

06:05:15.1 FIRST OFFICER: Comair 191 ready to go.

06:05:17.7 TOWER: Comair 191, Lexington tower, fly runway heading, cleared for takeoff.

06:05:21.0 CAPTAIN: Runway heading, cleared for takeoff, one ninety-one.

06:05:23.7 CAPTAIN: And line-up check.

06:05:25.1 TOWER: Eagle flight 882, that heading work for you? Do you wanna go, uh, north-west around the, uh, weather that's ahead of you?

06:05:30.7 EAGLE FLIGHT 882: Eagle 882, no that looks fantastic. Thank you very much.

06:05:34.4 CAPTAIN: Throw that bad boy on.

06:05:36.3 TOWER: Eagle flight 882, contact Indy Centre one two six point three seven. Good day.

06:05:39.5 EAGLE FLIGHT: Eagle 882 twenty-six, thirty-seven . . . eight eighty-two.

06:05:41.3 FIRST OFFICER: Transponder's on, packs on, bleeds closed, cleared for takeoff, runway heading [he does not call out number of takeoff runway, which is standard procedure. He is lined up on runway 26 and *not* runway 22, as he thinks he is and as he was instructed]. Six grand.

06:05:45.4 CAPTAIN: All right.

06:05:46.4 FIRST OFFICER: Anti-ice off, lights set, takeoff config's okay, line-up check's complete [again, no mention of runway number].

06:05:51.2 [Sound of clicks similar to pilot adjusting his seat]

06:05:57.6 CAPTAIN: All yours, Jim.

06:05:58.9 FIRST OFFICER: My brakes, my controls.

06:06:05.0 [Sound similar to increase in engine rpm]

06:06:07.8 FIRST OFFICER: Set thrust please.

06:06:11.7 CAPTAIN: Thrust set.

06:06:16.3 FIRST OFFICER: That is weird with no lights [on the sides of the runway, which should have been the first indication that they were taking off from the wrong runway].

06:06:18.0 CAPTAIN: Yeah.

06:06:24.2 FIRST OFFICER: One hundred knots.

06:06:25.1 FIRST OFFICER: Checks.

06:06:31.2 CAPTAIN: V one, rotate.

06:06:31.8 CAPTAIN: Whoa.

06:06:33.0 [Sound of impact]

06:06:33.3 [Unintelligible exclamation]

06:06:33.8 [Sound similar to stick shaker]

06:06:34.7 [Sound of chime]

06:06:34.7 [Sound similar to stall warning starts and continues to end of recording]

END OF RECORDING

The flight crew, instructed to take off from runway 22, had instead lined up the aeroplane on runway 26 before the takeoff roll. The aeroplane ran off the end of the runway and crashed into the airport perimeter fence, trees and terrain. The captain, flight attendant, and forty-seven passengers were killed. The first officer suffered serious injuries. The aeroplane was destroyed by impact forces and post-crash fire.

Index

Please see the following online glossaries for further explanations of technical terms:

http://www.freqofnature.com/aviation/aviation_glossary.html

http://www.gaservingamerica.org/library_pdfs/AVIATI_2.pdf